Cirencester College Library
Fosse Way Campus
Stroud Road
Cirencester
GL7 1XA

KT-495-870

A-LEVEL YEAR 2

STUDENT GUIDE

WJEC/Eduqas

Geography

Global governance: change and challenges

21st century challenges

Simon Oakes

Cirencester College, GL7 1XA
Telephone: 01285 640994
CIRCESTER COLLEGE LIBRARY
545954

HODDER
EDUCATION
AN HACHETTE UK COMPANY

Hodder Education, an Hachette UK company, Blenheim Court, George Street, Banbury, Oxfordshire OX16 5BH

Orders

Please contact Hachette UK Distribution, Hely Hutchinson Centre, Milton Road, Didcot, Oxfordshire, OX11 7HH

tel: 01235 827827

e-mail: education@hachette.co.uk

Lines are open 9.00 a.m.–5.00 p.m., Monday to Friday. You can also order through the Hodder Education website: www.hoddereducation.co.uk

© Simon Oakes 2017

ISBN 978-1-4718-6416-2

First printed 2017

Impression number 5 4 3

Year 2021

All rights reserved; no part of this publication may be reproduced, stored in a retrieval system, or transmitted, in any form or by any means, electronic, mechanical, photocopying, recording or otherwise without either the prior written permission of Hodder Education or a licence permitting restricted copying in the United Kingdom issued by the Copyright Licensing Agency Ltd, Saffron House, 6–10 Kirby Street, London EC1N 8TS.

This guide has been written specifically to support students preparing for the WJEC/Eduqas A-level Geography examinations. The content has been neither approved nor endorsed by WJEC/Eduqas and remains the sole responsibility of the author.

Cover photo: dabldy/Fotolia; other photos Simon Oakes

Typeset by Integra Software Services Pvt Ltd, Pondicherry, India

Printed in India

Hachette UK's policy is to use papers that are natural, renewable and recyclable products and made from wood grown in well-managed forests and other controlled sources. The logging and manufacturing processes are expected to conform to the environmental regulations of the country of origin.

Contents

Content Guidance

Global governance: change and challenges

Questions & Answers

■ Getting the most from this book

Exam tips

Advice on key points in the text to help you learn and recall content, avoid pitfalls, and polish your exam technique in order to boost your grade.

Knowledge check

Rapid-fire questions throughout the Content Guidance section to check your understanding.

Knowledge check answers

1 Turn to the back of the book for the Knowledge check answers.

Summaries

■ Each core topic is rounded off by a bullet-list summary for quick-check reference of what you need to know.

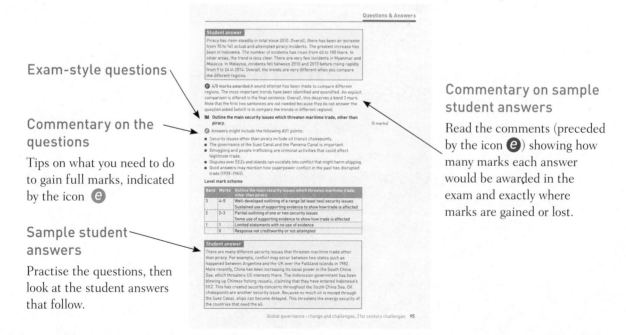

Exam-style questions

Commentary on the questions

Tips on what you need to do to gain full marks, indicated by the icon ⓔ

Sample student answers

Practise the questions, then look at the student answers that follow.

Commentary on sample student answers

Read the comments (preceded by the icon ⓔ) showing how many marks each answer would be awarded in the exam and exactly where marks are gained or lost.

■ About this book

This guide has been designed to help you succeed in the Eduqas and WJEC Geography A-level courses. The topics covered in this guide are:

■ Sections B and C of WJEC A-level Unit 3: Global governance: changes and challenges, and 21st century challenges
■ Sections B and C of Eduqas A-level Component 2: Global governance: changes and challenges, and 21st century challenges

The format of the different examination papers is summarised in the table below.

Specification and paper number	Total marks for sections B and C	Suggested time spent on sections B and C	Structured questions	Extended response/ essay
WJEC A2 Unit 3	61/96	75 min in paper lasting 2 h	Section B has *two* compulsory, structured questions with data response marked out of 17	Section B has an essay marked out of 18 (choose from *two*); section C has an essay marked out of 26 (choose from *two*)
Eduqas A-level Component 2	70/110	75 min in paper lasting 2 h	Section B has *two* compulsory, structured questions with data response marked out of 20	Section B has an essay marked out of 20 (choose from *two*); section C has an essay marked out of 30 (choose from *two*)

This guide has the following main objectives:

■ It provides you with key concepts, definitions, theories and examples that may be used to answer questions in the examination. The examples have been chosen to provide you with an up-to-date view of important global issues.
■ It provides guidance on how to tackle the synoptic element of the exam, which draws on ideas from your entire geography course.
■ It suggests self-study tasks that will enhance your knowledge and understanding before you enter the examinations.
■ Finally, it gives you the opportunity to test yourself through knowledge check questions, which are designed to help you to check your depth of knowledge. You will also benefit from noting the exam tips, which provide further help in determining how to learn key aspects of the course.

The section order follows the Eduqas specification.

The examinations

WJEC

The total examination time of WJEC A-level Unit 3: Global Systems and Global Governance is 2 hours. You have approximately 40–45 minutes to answer the questions in section B: **Global governance: change and challenges**, which comprises *two* compulsory structured, data response questions and *one* extended essay question. You then have approximately 30–35 minutes to answer the question in section C: **21st century challenges**, which comprises *one* compulsory extended essay question with resource material.

The two structured questions in section B will start with a part (a) resource-based question: you may be asked to describe a pattern or trend, or to suggest reasons why the data appear the way they do. The resource can also lead to a question requiring basic calculation or some interpretation of the data, which might be in cartographic, graphical, statistical or photographic form (including aerial and satellite images). The following part (b) question will develop from the topic in (a) and will require brief explanation of a relatively small section of the specification; it is worth 4 or 5 marks.

The section B essay is worth 18 marks. You will be expected to offer a structured response consisting of continuous prose. A command word or phrase such as 'discuss' or 'to what extent' will require you to adopt an evaluative approach and arrive at a conclusion. There will be a choice from two titles.

The section C essay is worth 26 marks. Once again, a command word such as 'discuss' will require you to adopt an evaluative approach and arrive at a conclusion. There will be a choice from two titles. Section C is supported by a suite of four resources — such as maps or photographs — which will relate to a range of overlapping ideas selected from both physical and human geography. Your purpose is to use this information along with your own recalled ideas in order to carry out a balanced evaluation of the question and arrive at a judgement.

Eduqas

At A-level, the total examination time of Component 2: Global Systems and Global Governance is 2 hours. You have approximately 40 minutes to answer the questions in section B: **Global governance: change and challenges**, which comprises *two* compulsory structured, data response questions and *one* extended essay question. You then have approximately 35 minutes to answer the question in section C: **21st century challenges**, which comprises *one* compulsory extended essay question (choose from two) with accompanying resource material.

The two structured questions in section B will start with a part (a) resource-based question worth 5 marks: you may be asked to describe a pattern or trend. The resource can also lead to a question requiring basic calculation or some interpretation of the data, which might be in cartographic, graphical, statistical or photographic form (including aerial and satellite images). The following part (b) question will develop from the topic in (a) and will require brief explanation of a relatively small section of the specification; it is worth 5 marks.

The section B essay is worth 20 marks. You will be expected to offer a structured response consisting of continuous prose. A command word or phrase such as 'discuss' or 'to what extent' will require you to adopt an evaluative approach and arrive at a conclusion. There will be a choice from two titles.

The section C essay is worth 30 marks. Once again, a command word such as 'discuss' will require you to adopt an evaluative approach and arrive at a conclusion. There will be a choice from two titles. Section C is supported by a suite of four resources — such as maps or photographs — which will relate to a range of overlapping ideas selected from both physical and human geography. Your purpose is to use this information along with your own recalled ideas in order to carry out a balanced evaluation of the question and arrive at a judgement.

How answers are marked

When your work is marked the examiner will be using **assessment objectives (AOs)**. The AOs for A2 (WJEC) and A-level (Eduqas) are as follows:

AO1: Demonstrate knowledge and understanding of places, environments, concepts, processes, interactions and change at a variety of scales.

AO2: Apply knowledge and understanding in different contexts to interpret, analyse and evaluate geographical information and issues.

- AO2.1a: Apply knowledge and understanding in different contexts to analyse geographical information and issues
- AO2.1b: Apply knowledge and understanding in different contexts to interpret geographical information and issues
- AO2.1c: Apply knowledge and understanding in different contexts to appraise/judge geographical information

AO3: Use a variety of relevant quantitative, qualitative and fieldwork skills:

- AO3.1: Investigate geographical questions and issues
- AO3.2: Interpret, analyse and evaluate data and evidence
- AO3.3: Construct arguments and draw conclusions

Mark bands

For AOs being tested in each question the marker will make use of marks bands for each AO to guide his/her decision. Here are the qualities that markers will be looking for in each mark band used to assess extended writing and essays.

Band 3

- Answers to extended writing questions (5 marks) will be clear, factually accurate and displaying good knowledge and understanding supported by developed examples, sketches and diagrams. Descriptions will be clear.
- Answers to **essays** (18, 20, 26 or 30 marks) will be well written, well structured and well argued. The command word (e.g. explain or evaluate) has been followed. Knowledge will be detailed, accurate and well supported by examples. Issues will be understood and thought about critically.

Band 2

- Answers to extended writing questions (5 marks) will often be unbalanced and partial; responses may be unstructured and making points in a random order. Knowledge is present but not always accurate or completely understood.
- Answers to **essays** (18, 20, 26 or 30 marks) will demonstrate some understanding but not all points and examples will be conveyed accurately. The style is more descriptive than discursive or evaluative. The range of ideas may be more narrow.

Band 1

- Answers to extended writing questions (5 marks) show only limited and fragmented factual knowledge. There may be no valid examples.
- Answers to **essays** (18, 20, 26 or 30 marks) may consist of unrelated and undeveloped ideas (possibly only in note form). The style may be entirely descriptive (with no

evaluation). The range of ideas may be narrow or the content may frequently lack relevance to the question being asked.

Geographical skills

You are expected to develop various skills as a geographer. Skills are both quantitative — using mathematical, computational and statistical procedures to record phenomena and processes — and qualitative — non-numerical techniques, using cartographic and GIS data, visual images, interviewing and oral histories. The specification provides a full list. Some statistical skills have been introduced in this guide and others will appear in the companion guides. The mathematical and statistical tests are not unique to this part of the course.

Specialised concepts

The following terms are essential for a twenty-first century A-level geographer to know and understand. Use them correctly in context whenever you can because the examination questions may expect you to show understanding of their meaning. They are of particular relevance to section C (21st century challenges), which may even incorporate a word such as **risk** or **resilience** as part of the essay title. Further exploration of some of these terms is included later on pp. 85–87.

Adaptation: the ability to respond to changing events and to reduce current and future vulnerability to change. States may need to adapt to a refugee crisis, for example.

Causality: the relationship between cause and effect. Everything has a cause or causes. Causes of migration include both push and pull factors; another cause is technology, which facilitates movement.

Equilibrium: a state of balance between inputs and outputs in a system. Steady-state equilibrium means there is balance in the long term but the system fluctuates in the short term. A country with a net migration balance of zero has an equilibrium of sorts.

Feedback: the way that environmental changes become accelerated, or are negated, by the processes operating in a human or physical system.

Globalisation: the process by which the world is becoming increasingly interconnected as a result of increased integration and interdependence of the global economy (see p. 10).

Identity: How people view changing places, landscapes or societies from different perspectives and experiences. The real or imagined characteristics that make somewhere or someone different from others (and which can affect migration).

Inequality: social and economic (income and wealth) inequalities between people and places. These inequalities give rise to movements of people.

Interdependence: relations of mutual dependence that are worldwide — states may become interdependent on account of migration and remittance flows (see p. 31).

Mitigation: the reduction of a phenomenon that is having a negative effect on people, places or the environment.

Place: a unique portion of geographic space. Places can be identified at a variety of scales, from local territories or locations to the national or state level. Places can be compared according to: their cultural or physical diversity; disparities in wealth or resource endowment; and the level of their local and global interactions with, or isolation from, other places.

Power: the ability to influence and affect change or equilibrium at different scales. Power is vested in citizens, governments, institutions and other stakeholders. Powerful countries are termed superpowers or regional powers.

Representation: how a country, place or area is portrayed by formal agencies (government and businesses) and informally by citizens, for example websites. It could affect migration rates towards a place.

Resilience: the ability of an object or a population to adapt to changes that have a negative impact upon them (see also p. 87).

Risk: the possibility of a negative outcome resulting from a physical process or human decision or course of action (see also p. 86).

Scale: places can be identified at a variety of geographic scales, from local territories to the national or state level. Global-scale interactions occur at a planetary level.

Sustainability: development that meets the needs of the present without compromising the ability of future generations to meet their own needs (see also p. 76).

System: a set of interrelated objects. A system can be either closed — with no import or export of materials or energy across its boundary — or open (where imports occur and are essential for system health).

Threshold: a critical limit or level that must not be crossed in order to prevent a system from undergoing accelerated and potentially irreversible change.

Content Guidance

Processes and patterns of global migration

■ Globalisation, migration and a shrinking world

Understanding globalisation

The umbrella term **globalisation** is used to describe a variety of ways in which places and people are increasingly connected with one another as part of a complicated global system. As Figure 1 shows, many widely differing definitions of the term globalisation are in use. Some emphasise that trade and the work of multinational corporations (MNCs) are at the heart of globalisation. Other definitions, as you can see, put greater emphasis on the cultural and political transformations that are also part of the globalisation process (Figure 2).

A rapid and huge increase in the amount of economic activity taking place across national boundaries has had an enormous impact on the lives of workers and their communities everywhere. The current form of globalisation, with the international rules and policies that underpin it, has brought poverty and hardship to millions of workers, particularly those in developing and transition countries. *UK Trade Union Congress*

Globalisation is a process enabling financial and investment markets to operate internationally, largely as a result of deregulation and improved communications. *Collins Dictionary*

The term 'globalization' refers to the increasing integration of economies around the world, particularly through the movement of goods, services, and capital across borders. It refers to an extension beyond national borders of the same market forces that have operated for centuries at all levels of human economic activity — village markets, urban industries, or financial centres. There are also broader cultural, political, and environmental dimensions of globalization. *IMF*

Globalisation

Globalization can be conceived as a set of processes which embodies a transformation in the spatial organization of social relations and transactions, expressed in transcontinental or interregional flows and networks of activity, interaction and power. *Held and McGrew (Globalization Theory)*

The expansion of global linkages, organization of social life on, global scale, and growth of global consciousness, hence consolidation of world society. *Frank Lechner (The Globalization Reader)*

A social process in which the constraints of geography on economic, political, social and cultural arrangements recede, in which people become increasingly aware that they are receding, and in which people act accordingly. *Malcolm Waters (Globalization)*

It might mean sitting in your living room in Estonia while communicating with a friend in Zimbabwe. It might mean taking a Bollywood dance class in London. Or it might be symbolized in eating Ecuadorian bananas in the European Union. *World Bank (for schools)*

Figure 1 Defining globalisation

Multinational corporations (MNCs) are businesses whose operations are spread across the world, operating in many nations as both makers and sellers of goods and services. Many of the largest are instantly recognisable 'global brands' that bring cultural change to the places where products are consumed.

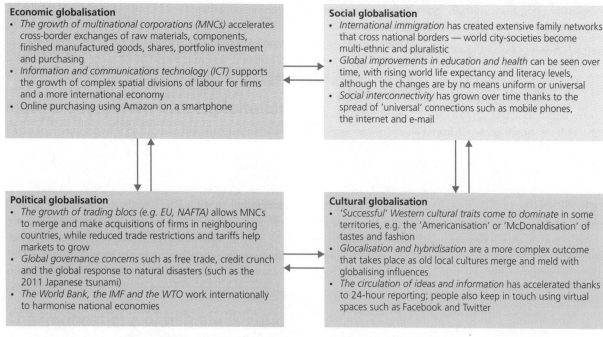

Figure 2 Four interconnected strands of globalisation

Modern globalisation is the continuation of a far older, ongoing economic and political project of global trade and empire building. Economies have been interdependent to some extent since the time of Earth's first great civilisations, such as ancient Egypt, Syria and Rome. There is nothing new about the global power-play and ambition of strong people, nations and businesses; one way of looking at globalisation is to see it as the latest chapter in a long story of globally connected people and places. In this guide, you will encounter both contemporary and more historical examples of international migration and global-scale trade and communications across oceans.

Modern globalisation *does* differ from the global economy that preceded it in three important ways. Over time, connections between people and places have:

■ **lengthened** — products are shipped greater distances than in the past; migrants and tourists travel longer distances from home

■ **deepened** — more aspects of everyday life have become globally connected: think about the food you eat each day and the many places it is sourced from via container shipping or air freight. It is difficult *not* to be connected to other people and places through the products we consume

■ **speeded up** — international migrants can travel quickly between continents using jet aircraft; they can also talk with their families at home in real-time, using technologies such as Skype

Is globalisation still accelerating or in retreat?

The sociologist Malcolm Waters defines globalisation as: 'A social process in which the constraints of geography on economic, political, social and cultural arrangements recede, in which people become increasingly aware that they are receding, and in which people act *accordingly*'.

- Following political events in 2016, it has become more important than ever to reflect critically on this idea of people acting 'accordingly'.
- In the UK and USA, many citizens have turned their backs on 'business as usual' globalisation.
- In both the US presidential election and the UK referendum on independence from the EU, migration and trade agreements were key issues affecting how people voted. In both cases, the victorious side campaigned in favour of reintroducing barriers to migration and trade. These are two key components of globalisation.

Global systems, connections and flows

In the study of global interactions, geographers conceptualise the world as consisting of **global networks** of connected places and people. A network is an illustration or model that shows how different places are linked together by connections or flows, such as trade or tourist movements. Network mapping differs from topographical mapping by not representing real distances or scale but instead focusing on the varied level of interconnectivity for different places, or nodes, positioned on the network map.

Network flows have stimulated the imagination of writers and artists, including Chris Gray who has represented parts of the world in the styles of the iconic London Underground network map (Figure 3).

Knowledge check 1

How globalised are you? Reflect on your own life and write a short essay examining how globalised you are. Use the framework provided in Figure 2 to structure your writing. Think critically about the extent to which you are a consumer of global commodities and global culture. Would you describe yourself as being a global citizen?

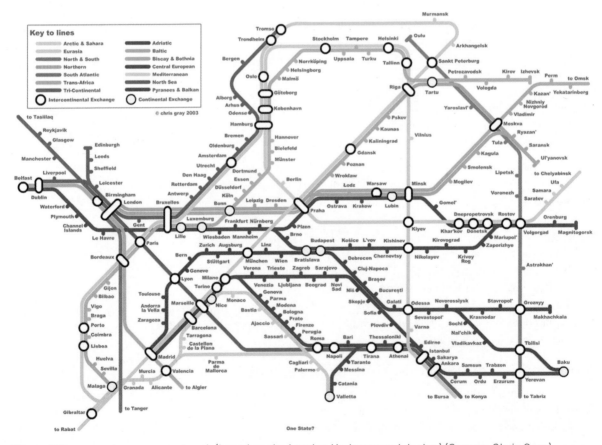

Figure 3 Europe redrawn as a network (based on the London Underground design) (Source: Chris Gray)

Table 1 Different kinds of global flow

Flows of food, resources and manufactured goods	■ In 2015, the value of world trade in food, resources and manufactured commodities exceeded US$25 trillion in value ■ One reason for this heightened activity is the rapid development of emerging economies especially China, India and Indonesia (combined, these countries are home to 3 billion people). Rising industrial demand for materials and increasing global middle class consumer demand for food, gas and petrol are responsible for almost all growth in resource consumption across nearly every category shown
Flows of money and financial services	■ In 2013, the volume of daily foreign exchange transactions reached US$5 trillion worldwide. Every day, huge capital flows are routed through stock markets in world cities such as London and Paris, where investment banks and pension funds buy and sell money in different currencies ■ Free-market liberalisation has played a major role in fostering international trade in financial services. For instance, the deregulation of the City of London in 1986 removed large amounts of 'red tape' and paved the way for London to become the world's leading global centre for financial services ■ Within the European Union, cross-border trade in financial services has expanded in the absence of barriers. Large banks and insurance companies are able to sell services to customers in each of the EU's member states ■ MNCs channel large flows of Foreign Direct Investment (FDI) towards the many different states they invest in
Flows of migrants and tourists	■ A record number of people migrated internationally in 2015, either for reasons of work or survival (sometimes the two are hard to distinguish) ■ The value of the international tourist trade doubled between 2005 and 2015. It is thought to be worth US$1 trillion annually (it is hard to make a precise estimate because of the many indirect benefits tourism creates). The number of international tourist arrivals doubled in the same period and now exceeds 1 billion people. Much of the new growth in touristic activity has been generated by movements within Asia ■ China now generates the highest volume of international tourism expenditure, while Europe receives more tourist arrivals than any other continent
Flows of technology and ideas	■ Global data flows have grown rapidly since the 1990s. Much recent expansion can be attributed to the growth of social media platforms and the arrival of on-demand media services ■ Faster broadband and powerful handheld computers have allowed companies such as Amazon and Netflix to stream films and music on demand directly to consumers

In Gray's worldview, the divisive international borders that separate states and cities are no longer present. Physical separation poses no obstacle to information flows between places in the internet age. The result is a borderless world of nodes and hubs, all connected by multi-coloured flow lines. These varied flows are analysed in Table 1.

Emerging economies are countries that have begun to experience higher rates of economic growth, often due to rapid factory expansion and industrialisation. Emerging economies correspond broadly with the World Bank's 'middle-income' group of countries and include China, India, Indonesia, Brazil, Mexico, Nigeria and South Africa.

The global middle class comprises people with discretionary income they can spend on consumer goods. Definitions vary: some organisations define the global middle class as people with an annual income of over US$10,000; others use a benchmark of US$10 per day income.

Exam tip

This part of the course includes a lot of important terminology such as globalisation and networks. Make sure you learn definitions carefully.

The global financial crisis (GFC) and global flows

The global flows in Table 1 have expanded enormously in size over time. However, growth has in some cases slackened or stalled since the **global financial crisis** (**GFC**) of 2007–09. The GFC originated in US and EU money markets, where sales of high-risk financial services and products valued at trillions of dollars eventually triggered the failure or near-collapse of several leading banks and institutions.

The resulting shockwaves undermined the entire world economy. The size of global gross domestic product (GDP) fell in 2009 for the first time since the end of the Second World War. Trade flows fell in value, many migrants returned home and fewer international tourist arrivals were recorded than in the previous year. Although global GDP has since begun to grow again, several key data indicators indicate that a cyclical or longer-term downturn in world trade flows has continued to affect developed, emerging and developing economies alike since the GFC.

- International flows of trade, services and finance grew steadily between 1990 and 2007 before collapsing and stagnating. 2016 was the fifth consecutive year when global trade did not grow; annual cross-border capital flows of US$3 trillion were well below their 2007 peak of US$8.5 trillion.
- Oil and some natural resource prices have fallen because of the global industrial slowdown. As a result, economic growth in sub-Saharan Africa halved between 2014 and 2016, leading several countries to ask the International Monetary Fund (IMF) for help.
- A significant slowdown of emerging economies has occurred: Brazil, Russia, South Africa and Nigeria recorded minimal growth or entered recession in 2016.

In contrast, global internet use and social networking have increased year-on-year, including the period 2007–09. For this reason, it is hard to arrive at an overall judgement about what is happening to global flows. Some flows are increasing while others have paused or are in retreat.

Classifying, quantifying and mapping migration

Globalisation has led to a rise in migration flows both *within* countries and *between* them.

- In 2013, 750 million **internal migrants** were residing in cities across the world (around one-third were Chinese rural–urban migrants). Global **urbanisation** passed the threshold of 50% in 2008, meaning that the majority of people now live in urban areas.
- Additionally, nearly one-quarter of a billion international migrants now live in countries other than the one in which they were born.

Classifying and quantifying international migrants

International economic migrants can be classified as belonging to one of two groups.

1 **Economic movers** are people who have moved voluntarily for reasons of work and the improved quality of life that higher earnings may bring. The overwhelming majority of movers, both at international and internal scales, are economic migrants.

Knowledge check 2

How do different types of global flow (food, migrants, money, ideas) give rise to different kinds of globalising process, such as cultural change, economic development or the spread of democracy?

Exam tip

If an exam question asks you to write about globalisation, structure your response by addressing different aspects of globalisation in turn (economic, social, cultural etc.)

An **internal migrant** is someone who moves from place to place inside the borders of a country. Globally, most internal migrants move from rural to urban areas ('rural–urban' migrants). In the developed world, however, people also move from urban to rural areas too (a process called counterurbanisation).

Urbanisation is an increase in the proportion of people living in urban areas.

2 **Refugees** are people who have been forced to leave their homes and to travel to another country. This group can be further subdivided into people fleeing conflict, political or religious persecution, or natural disasters, including drought and disease. Worldwide, there are now more than 20 million refugees.

It is important to recognise that the two groups overlap sometimes. Perspectives may vary on whether people escaping poverty in a drought-stricken farming region are best classified as refugees or voluntary migrants, for instance. Views could also differ on the point at which prejudice and persecution become so intolerable that people are compelled to leave their country. The distinction between voluntary and forced migration is important because people claiming refugee status must be able to prove their life was threatened in the country they abandoned.

Between 3% and 4% of the world's population are international migrants.

- The data used to calculate this percentage are drawn from various sources and in some cases are rough estimates only. Individual states have records of the numbers of foreign nationals who are resident. However, quantification is made difficult by the large volume of illegal migration flow that takes place. In some parts of the world, notably central Africa, large unchecked international flows of people occur in the absence of clear national boundaries and through a lack of policing or surveillance.

- It is believed that the *percentage* of the world's people who are migrants has not changed greatly over time despite the fact that the *number* of people migrating internationally has risen. This is because the total size of the world's population has grown too (between 1950 and 2015, world population rose from 4 billion to 7.3 billion).

Mapping the global pattern of migration

Important changes have taken place in the pattern of international migration in recent years.

- In the 1970s and 1980s, international migration was still directed mainly towards developed world destinations such as New York and Paris. The result was a **core–periphery** system: the developed economic core benefited from a **brain drain** of skilled workers from 'the global south' (including Indian doctors moving to join the UK's NHS). Since then, world cities in developing world countries such as Mumbai (India), Lagos (Nigeria), Dubai (UAE) and Riyadh (Saudi Arabia) have also begun to function as major global magnets for immigration.

- Much international migration remains relatively regionalised. In general, the largest labour flows connect neighbouring countries such as the USA and Mexico, or Poland and Germany.

Figure 4 shows how individual nations vary enormously in terms of the number or proportion of their population that is comprised of migrants. Partly this is explained by real or perceived economic opportunities these states offer potential migrants. Another important factor is differences in the level of each state's political engagement with the global economy. In order for a country to become deeply integrated into global systems, its government may need to adopt liberal immigration rules. Many of the UK's leading law firms have regional offices spanning the globe, from Singapore to New York. In order to maintain their global networks, these companies depend on foreign states granting UK lawyers permission to relocate permanently to their overseas offices.

Exam tip

In an essay, always seize an opportunity to show that some ideas (such as the distinction between forced and voluntary migrants) are not always clear-cut in the real world: demonstrating that you are aware of the imperfections in geographic theory provides evidence that you can think critically.

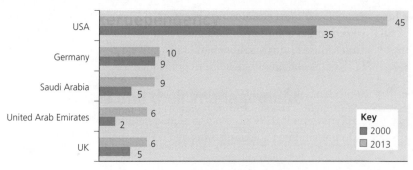

Figure 4 Number of international migrants living in selected countries, 2000 and 2013 (millions)

Figure 5 is a representation of the global pattern of international migration between 2005 and 2010. Important features that are clearly visible include:

- large volumes of **inter-regional** migration (most international migration originating in sub-Saharan Africa is directed towards other sub-Saharan nations)
- significant **intra-regional** flows linking North America with other regions including south Asia and central America

The single most important factor explaining the pattern shown is the uneven distribution of economic opportunity within global systems. As well as being triggered by economic inequality, migration also **reproduces** and perpetuates it sometimes. This is because the brain drain of talent away from source countries represents an economic loss that may only be partially offset by the receipt of **remittances** (money sent home).

Visualising migration flows

Movements of people are often shown as lines or arrows on a world map, but circular diagrams can be a more effective way of visualising global migration flows. This diagram shows migration between 196 countries between 2005 and 2010, broken down into fifteen world regions. The colour shows which region each flow came from, the width of a flow shows its size, and numbers indicate the total migration in and out of a region, in millions.

Figure 5 Circular flow diagram to illustrate global migration flows between 196 countries from 2005 to 2010

Self-study task 1

Apply your geographical knowledge and skills to the unusual representation of global migration flows shown in Figure 5. Make an estimate of the number of people (in millions) who migrated to North America from each Asian region between 2005 and 2010. As an extension task, discuss how Figure 5 is a representation of **interdependency** (p. 31)

A shrinking world for migration

Improvements in both the speed and capacity of transport and ICT (information and communications technology) are frequently cast as the key 'driver' of global migration. Important developments of the last 30 years — the internet and low-cost airlines, to name just two — have certainly helped accelerate population movements.

Heightened connectivity changes our conception of distance and potential barriers to the migration of people.

- This perceptual change has been described as time–space convergence (Janelle, 1968) and more recently as **time–space compression** (Harvey, 1990).
- Janelle plotted changing travel times between Edinburgh and London and found that a two-week stagecoach journey in 1658 was ultimately superseded by air flight lasting mere hours. He concluded that different places 'approach each other in space-time': they begin to feel closer together than in the past, as each successively improved transport technology chips away more minutes and hours from the connecting journey's duration.
- Since the sails of ships first filled with air, human society has experienced a **shrinking world**. As people's perception of space has changed — because of technology making the world feel smaller — long distances have ceased to be an intervening obstacle to migration (Figure 6).

An **intervening obstacle** is a physical, political or economic barrier that prevents migrants from completing their intended journey.

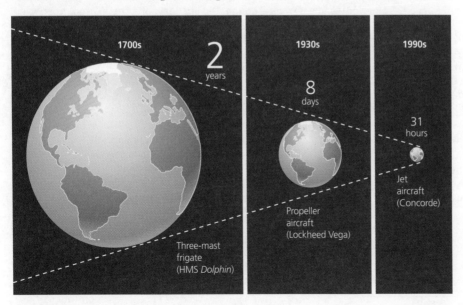

Figure 6 A shrinking world: the changing time taken to circumnavigate the world

Transport and communications developments

Historical studies of transport and communications are littered with innovation milestones stretching back over thousands of years. Table 2 shows four especially significant post-war technological and transport innovations that have helped increase interactions between places in ways which have fostered international migration.

Table 2 Transport and communications development have led to a shrinking world for migrants

Mobile phone	Lack of communications used to be a big obstacle to international migration. Without connectivity, people will not know 'the grass is greener' elsewhere and are more likely to stay where they are. In recent years, however, global telecommunications growth rates have reached extraordinary levels (Figure 7). ■ In 2005, 6% of all Africans owned a mobile phone. By 2015 this had risen ten-fold to 60% because of falling prices and the growth of provider companies such as Kenya's Safaricom. Only 10% of Africa's population live in areas where no mobile service is available ■ Rising uptake in Asia (in India, more than 1 billion people are mobile subscribers) means there are now more mobile phones than people on the planet
Internet	Increasingly, many mobile phones provide internet access too. Never before has it been easier for potential migrants to find out about the opportunities other countries offer. ■ International migrants increasingly communicate with one another using smartphone apps. In 2015, a Facebook group called 'stations of the forced wanderers' helped more than 100,000 migrants to exchange advice on how to avoid authorities and find routes across European borders using GPS information ■ The **media representation** of places may also affect people's decision to migrate. Film clips on YouTube portray life in some countries very positively, prompting more people to move there ■ Migration becomes easier when people can maintain long-distance social relationships more easily than in the past using the internet. Since 2003, Skype has provided a cheap and powerful way for migrants to maintain strong links with family they have left behind ■ The militant political group Daesh (or so-called IS) has used social media and YouTube to encourage young and men and women from the UK, France and other countries to migrate to the Middle East to join its ranks
Air travel	Air travel has become more affordable over time, allowing more people to move internationally. ■ The introduction of the intercontinental Boeing 747 in the 1960s made international travel more commonplace ■ Recent expansion of the cheap flights sector, including easyJet, has brought it to the masses in Europe: most of Europe's major cities are now interconnected via easyJet's cheap flight network (with 65 million passenger flights in 2014) ■ The growth of the global middle class has driven the expansion of flights between Asian countries too; East African Safari Air Express caters for higher earners in Kenya and its neighbours
High-speed rail	Railways remain an important means of travel between neighbouring countries. ■ Railways are the chief conduit linking rural and urban parts of China. Migrant workers travel in both directions along the route of the 1,500 km China–Tibet 'sky train', whose hi-tech specifications can survive the Tibetan plateau where temperatures drop to −35°C

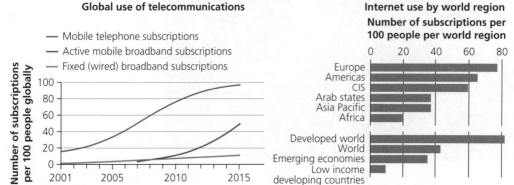

Figure 7 Global trends and patterns for communication technology use

'Switched-off' people and places

It remains the case that not everyone has access to communications technology. Neither does everyone have the money or political freedom needed to travel internationally.

■ The geographer Doreen Massey wrote critically about changing perceptions of place in a technologically advancing world. She argued that time–space compression is socially differentiated: not everyone experiences the sense of a shrinking world to anything like the same extent because of income differences.

■ Billions of people still cannot afford the cost of a smartphone and broadband subscription.

■ Political factors also play a role in the persistence of a digital divide between 'switched-on' and 'switched-off' societies (such as North Korea).

Summary

■ Globalisation is a complex set of processes that includes international migration along with global trade, data and money movements.

■ Globalisation has accelerated over time. However, some (but not all) global flows have slowed recently.

■ Voluntary and forced international migration flows have grown in volume over time. Important global patterns can be identified that connect core and peripheral regions of the global economy.

■ New technologies play a crucial role in enabling migration. However, not all people have access because of political factors and poverty.

Knowledge check 3

Think about how the shrinking world process has led to increased migration between countries. In your view, which technology has had the greatest impact? What evidence might help support your judgement?

■ Causes of international economic migration

Economic push factors driving international out-migration

What factors explain the large volumes of international migration that take place every year? What kind of imbalances and injustices in the global economic system are responsible for generating such large-scale movements? Recurring structural causes of international migration include: **poverty**, **primary commodity prices** and **poor access to markets** within global systems.

Poverty

The world can be divided crudely into a **core** of developed countries, a **semi-periphery** comprised of emerging economies and a **periphery** of low-income developing countries. The characteristics of these three global groups are outlined in Table 3.

Table 3 The three main groups of countries classified by per capita income and economic dynamics

Global periphery Low-income developing countries (LICs)	Around 30 countries are classified by the World Bank as having low average incomes (GNI per capita) of US$1,045 or below (2015 values). Agriculture plays a key role in the economies of LICs.
Semi-periphery Emerging economies (EEs)	There are around 80 countries currently experiencing higher rates of economic growth than in the past, usually on account of rapid factory expansion and industrialisation. The EEs correspond with the World Bank's 'middle-income' group of countries.
Global core High-income developed countries (HICs)	This group of around 80 countries is classified by the World Bank as having high average incomes of US$12,736 or above (2015 values). Office and retail work has overtaken factory employment, creating a post-industrial economy. Around half of all HICs are very small countries, including Bahrain, Qatar and the Cayman Islands.

This threefold division of the world generates the economic movements of people shown in Figure 5 (p. 16). As part of this analysis, it is important to distinguish between different types of poverty as migratory triggers.

■ Movements from the world's poorest countries are rooted in the **extreme poverty** of the source region. People living below the World Bank's US$1.90 extreme poverty line are often unable to meet their own basic needs of food, clothing and shelter. Inevitably, these conditions lead to out-migration even if people are moving no further than neighbouring countries, where conditions may be little better.

■ Movements from emerging economies such as Poland and Mexico to developed countries such as the UK and USA can be explained in terms of the **relative poverty** of the source regions. When Poland joined the EU in 2004, its gross domestic product (GDP) per capita was just US$12,600, around two-fifths that of the UK. While Poland's GDP per capita was high by sub-Saharan African standards, it was sufficiently low to prompt an estimated 1 million young Poles to seek a better life in the UK.

Primary commodity prices

In theory, a country's primary commodities (unprocessed food, timber, minerals and energy resources) provide it with the opportunity to trade with other countries, thereby generating the income needed for economic development to take place. In reality, this does not always happen. Countries that trade only in agricultural produce and raw materials do not always gain a good income. This means they have insufficient money to import expensive manufactured products from developed countries. Development goals become harder to achieve without computers for schools or specialised hospital equipment. As a result of persisting poverty, out-migration of skilled, ambitious and talented people can rob a country of its most valuable human resources, thereby creating even greater development challenges through the process of **positive feedback**.

There are two key reasons why prices paid for primary commodities in global markets are often low.

1 **Overproduction** occurs when too many countries grow the same crop. This oversupply pushes down prices globally. When crop yields are especially high

Exam tip

Make sure you do not write in an over-general way about world development. Nobody should be describing the world as consisting of 'MEDCs and LEDCs' any more: times have changed.

because of good weather, the problem worsens. In some years, prices for coffee beans, cocoa beans or bananas have fallen very low, bringing misery to producer communities.

2 **Poor governance** is a major contributing factor. In the past, developing countries sometimes lacked the human capital needed to strike good trade deals: when Democratic Republic of the Congo (DRC) gained independence from Belgium in 1960, the country was reputed to have only 16 citizens with degrees. Foreign mining, timber and food companies allegedly took advantage of the widespread lack of business and economic expertise in DRC by brokering deals to buy the country's resources at a fraction of their real market value. In later years, the exploitation continued with the willing collaboration of DRC's President Mobutu.

There are exceptions. Prices paid for oil, diamonds, gold and rare earths can be very high. However, the presence of these commodities sometimes creates an even greater developmental challenge by giving rise to conflict. This is the **resource curse theory**. DRC's metals, Sierra Leone's diamonds and South Sudan's oil supplies have often made life worse and not better for citizens. Conflict refugees have joined the economic migrants leaving these countries.

Poor access to markets within global systems

The division of the world into large trading areas called **trade blocs** is another reason why poverty persists in some developing countries.

■ The European Union (EU) is a trading area that protects its own farmers by placing **import tariffs** on food imports from other countries. As a result, farmers in non-EU countries such as Kenya find it harder to get a good price for the food they sell to European supermarkets.

■ In addition, high levels of government financial support allow European farmers to produce meat and vegetables cheaply. As a result, African farmers must offer to sell their own produce at even lower prices to firms such as Aldi if they want to trade.

The **World Trade Organization** (**WTO**) aims to reduce unfair trade barriers and government subsidies globally. However, attempts to persuade developed countries to abandon agricultural protectionism and improve market access for poorer countries have not always succeeded.

Cultural and political drivers of international migration

Sometimes international migration is encouraged by particular cultural and political factors. These include the existence of diaspora communities, **post-colonial** links between states and political legislation permitting **free movement**.

Global diaspora communities

Global diasporas provide important contexts or frameworks for international migration: it makes sense to migrate to a country where large numbers of one's fellow citizens are already living. Joining an established diaspora community offers many benefits: it may make finding work easier; family members may provide support during the potentially difficult initial period of relocating and settling in.

Diaspora is the worldwide scattering or dispersal of a particular nation's migrant population and their descendants.

Several famous diaspora examples are shown in Table 4 and Figure 8. Other notable examples include the French, Italian, Mexican, Brazilian, Nigerian and Malay diasporas. The UK's 'Celtic fringes' have all given rise to significant global diasporas despite these nations' relatively small population sizes. For instance, Ireland is home to just 4 million people, yet over 70 million individuals living worldwide claim Irish ancestry. In the USA alone, 30 million people believe themselves tied to Irish bloodlines. The Welsh diaspora includes communities in Patagonia (South America) who first arrived there in 1865.

Table 4 Examples of global diaspora populations

The Chinese diaspora	The neighbouring countries of Indonesia, Thailand and Malaysia, along with far-flung places such as the UK and France, have significant Chinese populations. In many world cities, clearly delimited 'Chinatown' districts exist. A thousand years of sea-faring trade gives this diaspora a long history. The arrival of Chinese MNCs in Africa has brought further diaspora growth in recent years.
The Indian diaspora	This is one of the world's largest, numbering 28 million in 2016. People of Indian citizenship or descent live in almost every part of the world. Important features of the pattern are that it numbers more than 1 million in each of 11 countries. The largest concentrations are in the USA, the UK, Malaysia, Sri Lanka, South Africa and the Middle East.
The 'Black Atlantic' diaspora	This has been described by Paul Gilroy as a 'transnational culture' built on the movements of people of African descent to Europe, the Caribbean and the Americas. A shared, spatially dislocated history of slavery originally helped shape this group's identity. Today, international connectivity is maintained through migration, tourism and cultural exchanges across the Atlantic that are well exemplified by an international Black music scene that has given the world jazz, blues, reggae and hip-hop.
The Scottish diaspora	Like Ireland, Scotland is a small country of only a few million residents yet has a diaspora numbering tens of millions. Online ancestry websites enable people living all over the world to trace their roots back to Scotland: this is another interesting way in which technology has influenced global migration. People who discover they have roots in another country may become more likely to consider moving there. GlobalScot is a website run by government-funded Scottish Enterprise that actively encourages members of the Scottish diaspora to network economically with one another.

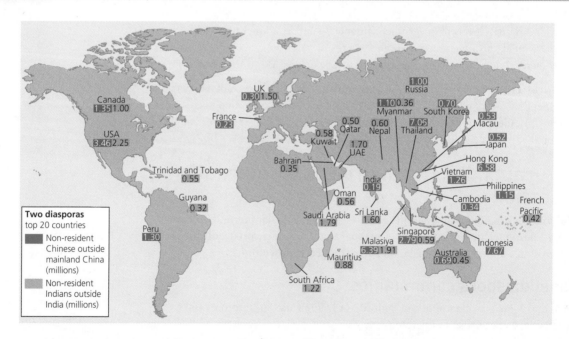

Figure 8 Non-resident Indian and Chinese citizens living abroad form part of both countries' diaspora (millions more people of Indian or Chinese descent are also scattered world wide) (Sources: CEIC, OCAC, MOIA, US Census Bureau)

Post-colonial movement of people

Between the 1950s and 1970s, the UK received migrants from the Caribbean (especially Jamaica), India, Pakistan, Bangladesh and Uganda. Smaller numbers came from Nigeria, Kenya and other ex-British African territories. Today, these countries are **Commonwealth** members.

- Migrants came to fill specific gaps in the labour force that opened up after the Second World War. Sometimes, migrants were recruited directly (London Underground held interviews for bus drivers in Kingston, Jamaica).
- There was still a large demand for workers in heavy and light industry, especially in the textile mills of the Midlands, Lancashire and Yorkshire.
- There were gaps in the skilled labour market too, notably within the ranks of the new National Health Service (UK doctors had not been trained in sufficient numbers during the 1930s and 1940s to fully staff the ambitious new NHS). Many doctors travelled to the UK from India, Pakistan and parts of Africa.

Other European countries that were once great world powers have post-colonial ties with states that once belonged to their own empires. France is home to the descendants of many Algerian and Tunisian migrants; some Belgian citizens and residents migrated there from DRC.

Rules permitting free movement in the EU, South America and Africa

Within the EU, free movement of labour is permitted. Southern England, northern France, Belgium and much of western Germany are important host regions for much of the migration that has occurred. This area includes the world cities of London, Paris, Brussels and Frankfurt. Labour migration flows from source regions in eastern and southern Europe have been overwhelmingly directed towards these places. Most national border controls within the EU were removed in 1995 when the **Schengen Agreement** was implemented. This enables easier movement of people and goods within the EU, and means that passports do not usually have to be shown at borders.

Other world regions have begun to adopt free movement rules too.

- South American countries have also taken steps towards this goal. Between 2004 and 2013, nearly 2 million South Americans obtained a temporary residence permit in one of nine countries implementing the agreement. After the signing of the **Mercosur Residency Agreement**, nationals of Argentina, Bolivia, Chile, Colombia, Peru, Uruguay and Venezuela have the right to apply for temporary residency in another member state. After two initial years of temporary residency, it is possible to convert the temporary status to permanent residency.
- The **African Union** has said it wants to break down borders through closer integration. In 2016, the African Union (which has 54 member states) began issuing e-passports that permit recipients to enjoy visa-free travel between member states.

Knowledge check 4

How much personal experience do you have of migration? Do you or any of your friends belong to a diaspora? Can you identify famous people in UK sports, television or the media who belong to a diaspora community?

Understanding superpowers

The term **global superpower** was used originally to describe the ability of the USA, USSR (Russia) and the UK to project power and influence anywhere on Earth to become a dominant worldwide force. A **regional superpower** exerts significant influence over its neighbours: Nigeria, South Africa and Australia are good examples.

Global superpowers

- The UK was a colonial power, alongside France, Spain, Portugal, Italy, Belgium and Germany. Between approximately 1500 and 1900, these leading powers built global empires. One result was the diffusion of European languages, religions, laws, customs, arts and sports on a global scale.

- In contrast to the direct rule of the British in the 1800s, the USA has dominated world affairs since 1945 mainly by using indirect forms of influence or 'neocolonial' strategies. These include the US government's provision of international aid and the cultural influence of American media companies (including Hollywood and Facebook). Alongside such **soft power** strategies, the US government has routinely made use of **hard power**. This means the geopolitical use of military force (or the threat of its use) and the economic influence achieved through forceful trade policies including economic sanctions or the introduction of import tariffs. The term **smart power** is used to describe the skilful combined use of both hard and soft power in international relations (Figure 9).

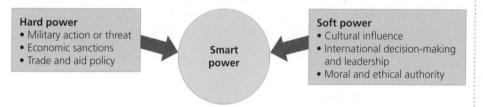

Figure 9 The ingredients of smart power

Other than the USA, which other states can claim currently to be true global superpowers?

- China became the world's largest economy in 2014 by one measure, and exerts great influence over the global economic system through its sheer size. Table 5 compares the USA and China.

- Although no single European country can equal the influence of the USA any more, several have remained significant global players in the post-colonial world (notably the G8 nations of Germany, France, Italy and the UK). Another view is that European states can only rival the USA's global superpower status when they work together as members of the European Union.

The political scientist Joseph Nye coined the term **soft power** to mean the power of persuasion. Some countries are able to make others follow their lead by making their policies attractive and appealing. A country's culture (arts, music, cinema) may be viewed favourably by people in other countries.

Hard power means getting your own way by using force. Invasions, war and conflict are blunt instruments. Economic power can be used as a form of hard power: sanctions and trade barriers can cause great harm to other states.

Table 5 Analysing and evaluating the superpower status and global influence of the USA and China

	Analysis	Evaluation
USA	■ The 320 million people who live here (less than one-twentieth of the world's population) own more than 40% of global personal wealth. Of the 500 largest global companies, one-quarter were US-owned in 2015. ■ US cultural influence is so strong that terms such as 'Americanisation' and 'McDonaldisation' are widely used to describe the way American food, fashion and media have shaped global culture. No wonder many people want to migrate there. ■ The USA has used military power and covert intelligence operations to intervene in the affairs of almost 50 states since 1945.	■ The USA's influence over international organisations, including the UN, NATO, the IMF and the World Bank, has given it greater influence over global politics than any other state. The USA was the main architect of the global economic system created at the end of the Second World War. The economic principles that underpin globalisation have become known as 'the Washington consensus'. ■ The USA is a true global superpower. No other country has such a formidable combination of geopolitical, economic and cultural tools at its disposal.
China	■ China's growth began in 1978 when Deng Xiaoping began the radical 'open door' reforms that allowed China to embrace globalisation while remaining under one-party rule. ■ Today, China is the world's largest economy. Over 400 million of its people are thought to have escaped poverty since the reforms began. FDI from China and its MNCs is predicted to total US$1.25 trillion between 2015 and 2025.	■ The average income of China's population is still less than one-third that of US citizens. Recently, its economic growth has slowed ■ China lacks the soft power of the USA, in part because of its cultural isolation from the rest of the world (few foreign films are allowed into China and internet freedoms are restricted). The lack of democracy in China also affects it relations with some other countries adversely.

Regional superpowers

Many countries can make a claim to be regionally powerful and to also exert global influence in certain ways. For example, the tiny Middle Eastern state of Qatar has the highest per capita GDP in the world, in excess of US$100,000. Its wealth and global influence, like neighbouring Saudi Arabia's, is partly due to fossil fuel sales: Qatar has 14% of all known gas reserves. Qatar's government has re-invested its **petrodollar wealth** in ways that have diversified the national economy and built global influence too.

■ The city of Doha has become a powerful place where international conferences and sporting events are held, served by Qatar Airways and Doha International airport. Important United Nations and World Trade Organization (WTO) meetings have taken place in Doha, including the 2012 United Nations Framework Convention on Climate Change (UNFCCC) climate negotiations. The city is set to host the 2022 football World Cup.

■ Qatar's Al Jazeera media network rivals the BBC and CNN for influence in some parts of the world and is an important source of soft power.

■ However, many people regard Qatar as a regional power, rather than a true global superpower.

The benefits of international migration for superpower states

At certain moments in their history, superpower states have encouraged immigration as part of a carefully calculated growth strategy. There are two kinds of labour shortage that countries may experience from time to time:

1 **Skilled labour shortages:** a shortage of people trained in a particularly important profession, such as medicine or engineering, can be economically or socially damaging for a country.

> **Knowledge check 5**
>
> Make a case for categorising the following countries as being either regional or potentially global superpowers: Nigeria, South Africa, India, Russia, Brazil, Japan and Saudi Arabia. Which criteria matter most when thinking about this?

2 **Unskilled labour shortages:** too few people willing to do poorly paid but essential work, such as building construction, can impede economic progress.

As we have already seen, the British state actively encouraged post-colonial migration to take place when it found itself desperately short of labour after the Second World War in 1945 (Figure 10). This consisted of the voluntary movement of people from former colonies of the British Empire. Some worked in factories; skilled individuals filled important medical posts in the newly formed NHS. One reason why it was so easy for the UK to achieve this aim was because of its language, customs and traditions. These had been introduced to British territories under colonial rule in the 1800s. Medical schools in India used the same textbooks as British teaching hospitals. The populations of ex-colonies spoke fluent English and showed an affinity with the British way of life. The UK was therefore able to take advantage of its past influence over these countries by advertising work opportunities to young Asians and Africans who were excited to move to the UK, following an education in schools where British history and culture would have been celebrated.

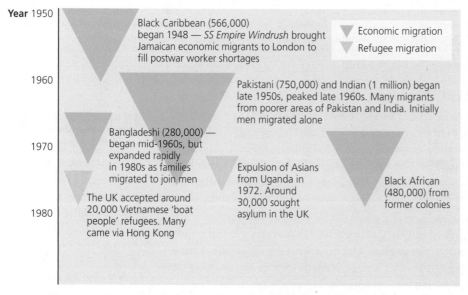

Figure 10 UK post-colonial migration

Table 6 shows how the USA and Australia have benefited from sometimes selective in-migration of people from other states. However, it is not the case that all powerful states are always in favour of allowing immigration to take place. Less than 2% of the Japanese population is foreign or foreign-born. Despite Japan's important status as a global economic power, migration rules have made it tough for newcomers to settle permanently. Nationality law makes the acquisition of Japanese citizenship by resident foreigners an elusive goal (the long-term pass-or-go-home test has a success rate of less than 1%). Japan faces the challenge of an **ageing population**, however. There will be three workers for every two retirees by 2060. Many people think that Japan's government must relax its attitude towards immigration if it wants to maintain its economic role in the world.

Table 6 How the USA and Australia have benefited over time from immigration

USA	<ul><li>The USA is home to around 300 million people of foreign ancestry who share the territory with just 3 million native Americans. In this 'immigrant nation', we find people of Scottish, Irish, Italian, Greek, Jamaican, Puerto Rican, Indian, Swedish, Polish, Jewish, German, Korean, Nigerian, Jamaican and Chinese descent, among many others. The country has prospered over time precisely because it has attracted so many young and talented migrants.</li><li>Since the 1990s, skilled Indian migrants have travelled to the USA in large numbers. The country's Indian diaspora community is more skilled and highly paid than any other US migrant community.</li><li>Many of the USA's most famous cultural exports — and sources of its soft power — are in fact derived from the culture of immigrant groups. African-American-influenced rock and hip-hop music are among these; so too are the world-famous American hamburger and apple pie (which are thought to be German and Dutch in origin).</li></ul>
Australia	<ul><li>Australia's government uses migration policies carefully to maintain strength in any economic sector where labour shortages are evident. The country currently operates a points system for economic migrants called the Migration Programme. In 2013, only 190,000 economic migrants were granted access to Australia (this figure included the dependants of skilled foreign workers already living there). The top five source countries were India, China, the UK, the Philippines and Pakistan.</li></ul>

Global hubs

Superpower demand for migrant labour is often concentrated in particular **global hubs** within these states.

A global hub is a particularly important city when viewed *at both a national and global scale*. This is on account of the presence of the headquarters of major MNCs, globally renowned universities, global financial or political institutions or other world-class assets. Global hubs such as New York and Mumbai have gained in economic strength over time by attracting flows of foreign investment and the international workforce this brings. There are large numbers of foreign workers in London's Canary Wharf financial district. They play a vital role managing the European operations of American, Chinese, Indian, Japanese and Singaporean companies who have established offices there.

- Some global hubs are **megacities** with more than 10 million residents. Size is not a prerequisite for global influence, however. Smaller-sized global hubs which 'punch above their weight' in terms of their global reach include Washington DC and Qatar's Doha, as we have already seen. In 2016, Oxford University in the UK was named as the world's leading educational institution: despite its small size, the city of Oxford is a powerful place. It attracts many overseas migrant students and lecturers.

- Figure 11 shows how global hubs grow in power and influence over time from an often advantageous pairing of natural resources and human resources. International migration plays a vital role in continually replenishing the human resources of powerful cities and states.

Knowledge check 6

Develop an argument about the importance of the role played by global migration flows in helping superpower countries to gain and maintain their status. Don't forget to consider Japan as one of your examples.

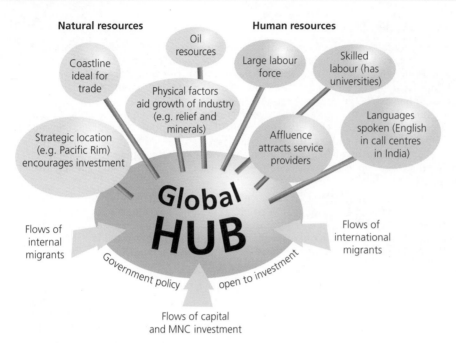

Figure 11 Migration plays a vital role in feeding the growth and continued prosperity of global hubs

Superpower out-migration

In addition to the role played by in-migration, we must not forget that the out-migration of citizens also helps superpowers to gain worldwide influence. Large numbers of UK and US citizens have moved overseas; Indian and Chinese diasporas in African have grown significantly in size recently. In each of these cases, diaspora members have the capacity to act as unofficial 'soft power ambassadors' for their country of origin or ancestry.

Summary

- There are important structural causes of economic migration taking place within the world economic system. Economic inequalities and injustices continue to drive large movements of people from state to state.
- History and culture shape the global pattern of voluntary labour movements. Links were created between places in the past under the colonial system; these connections continue to influence present-day movements of people.
- Political factors and migration rules play a major role in determining who is free to move and who is not.

- Some states have become global superpowers, or major regional powers, on account of the economic, military, political or 'soft' cultural influence they exert over other countries.
- Powerful countries such as the USA and UK are major magnets for international migration. They use other countries' human resources in ways that help them reproduce their own economic strength and influence over time.

■ Managing the global consequences of international economic migration

Migration flows and economic inequalities

The arrival of large numbers of low-skilled workers in a country can result in sizeable remittance flows directed towards the source country. Low-waged international migrants are drawn towards global hubs in large numbers. London, Los Angeles, Dubai and Riyadh are all home to large numbers of legal and illegal immigrants working for low pay in kitchens, on construction sites or as domestic cleaners. Around US$500 billion of remittances are currently sent home by migrants annually (Figure 12). This is three times the value of overseas development aid.

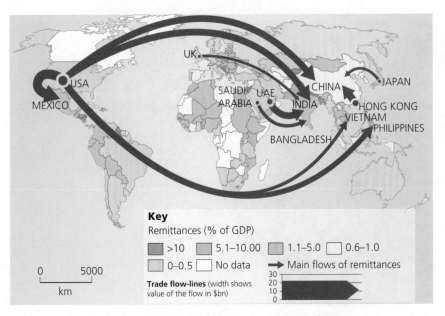

Figure 12 Flows of migrant remittances, 2011

These large-scale migrations also lead to significant cultural change in the host country, such as increased diversity of religious ideas. For instance the British Indian Sikh community is now the largest Sikh community outside India — Gurdwara temples can be found in Cardiff and other major cities. An exchange of ideas can also be seen in the impact young British Asians have had on popular culture: London's Jay Sean (whose real name is Kamaljeet Singh Jhooti) has brought elements of south Asian music into Britain's mainstream.

The movement of smaller numbers of high-skilled workers and high-wealth individuals also has significant effects for both source and host country alike. The variety of these **elite migrants** (highly skilled and/or socially influential individuals) is shown in Figure 13. Their wealth derives from their profession or inherited assets. Some elite migrants live as 'global citizens' and have multiple homes in different countries. They encounter few obstacles when moving between countries. Most governments welcome highly skilled and extremely wealthy migrants.

As Figure 13 shows, many elite migrants work in the knowledge economy, including writers, musicians and software designers. Skilled IT professionals from the USA, India and elsewhere work in the UK's quaternary industry clusters in cities such as Bristol, London and Cambridge.

■ This benefits the UK as a host nation for migration.
■ India, however, is sometimes said to suffer from a brain drain. This is on account of the large numbers of medical and IT professionals who have migrated away from this source nation and towards other states.

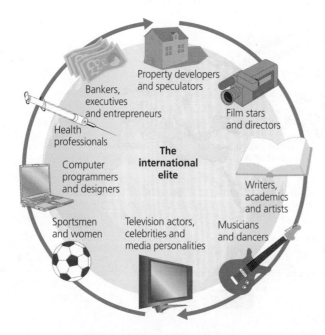

Figure 13 Global elite migration

Table 7 illustrates how selected migratory movements and their associated flows of money and ideas may:
■ reduce global inequalities and promote growth and stability
■ exacerbate global inequalities and potentially lead to tension or conflict

Table 7 The consequences of selected migratory movements and their associated flows of money and ideas

	Host regions	Source regions
Positive effects (reducing inequality and promoting growth)	■ Fill particular skills shortages (e.g. Indian doctors arriving in the UK in the 1950s). ■ Economic migrants willingly do labouring work that locals may be reluctant to (e.g. Polish workers on farms around Peterborough). ■ Working migrants spend their wages on rent, benefiting landlords, and pay tax on legal earnings. ■ Some migrants are ambitious entrepreneurs who establish new businesses employing others (in 2013, 14% of UK start-up businesses were migrant-owned).	■ In Bangladesh, the value of remittances exceeds foreign investment. Unlike international aid and lending, remittances are a peer-to-peer financial flow: money travels more or less directly from one family member to another. This money flow helps the social development of communities that have previously been excluded financially from access to education and healthcare. ■ In time, migrants or their children may return, bringing new skills (young British Asians have relocated to India to start health clubs and restaurant chains).
Negative effects (exacerbating inequality and promoting tension)	■ Social tensions arise if citizens of the host country believe migration has led to a lack of jobs or affordable housing (a view adopted by some UK newspapers). ■ Local shortages of primary school places due to natural increase among a youthful migrant community (e.g. London Boroughs that have become eastern European migration 'hotpots'). This places a financial burden on local authorities. ■ Employers may favour using migrants instead of 'native' workers; working-class communities may suffer from unemployment as a result.	■ The economic loss of a generation of human resources, schooled at government expense, including key workers such as teachers and computer programmers. Poland has lost young people every year since the 1960s ■ The Increase in the proportion of aged dependents creates a long-term economic challenge. ■ There is reduced economic growth as consumption falls (especially urban services and entertainment for a young adult market: many nightclubs closed in Warsaw in 2004 when Poland joined the EU). ■ There is no guarantee that remittances will continue to be sent home in the long-term.

Migration and interdependency

Over time, global flows have created networks of interconnected and **interdependent** places. Every country depends to some extent on the economic health of others for its own continued wellbeing. Interdependency may have economic, social, political and environmental dimensions (Figure 14).

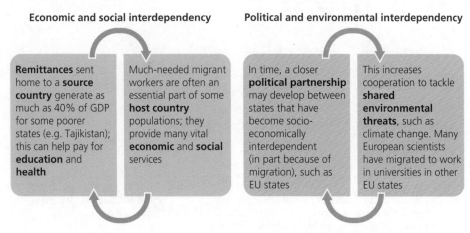

Figure 14 How migration can foster increased economic, social, political and environmental interdependency between states

Examples of interdependency

When large-scale migrant labour flows become focused on **core** or hub regions, a process called **backwash** is said to be taking place. This process can drain **peripheral** places of young workers.

- Within the EU, the Schengen Agreement (p. 23) has accelerated the backwash process at an international level (Figure 15).
- Many economists believe backwash works in everyone's interest. Migration is viewed as an efficient way of making sure that economic output is optimised for the EU as a whole.
- In turn, this provides EU governments with greater tax revenues to pay for services and infrastructure, which are shared with all member states, including road-building projects, payments for farmers and grants for new businesses. According to core–periphery theorist John Friedmann, backwash effects are being balanced out by the **trickle down** of wealth to every country. The result is a truly interdependent alliance of states.
- Critics of this **neoliberal** model argue that backwash migration 'losses' for peripheral states in Eastern Europe are really far greater than any trickle-down 'gains' they may experience. In reality, it is hard to either accept or reject the hypothesis because of the sheer complexity of the economic and demographic processes involved.

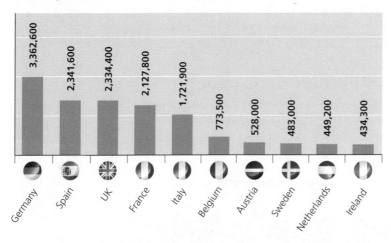

Figure 15 The 10 EU states with the largest proportion of people born in another EU state, 2012 (actual numbers shown)

Table 8 shows further examples of interdependency between states arising on account of international migration.

Table 8 How migration and remittance flows create economic interdependency

Indian workers moving to UAE	Over 2 million Indian migrants live in United Arab Emirates, making up 30% of the total population. Many live in Abu Dhabi and Dubai. An estimated US$15 billion is returned to India annually as remittances. Most migrants work in transport, construction and manufacturing industries. Around one-fifth are professionals working in service industries.
Filipino workers moving to Saudi Arabia	Around 1.5 million migrants from the Philippines have arrived in Saudi Arabia since 1973, when rising oil prices first began to bring enormous wealth to the country. Some work in construction and transport industries, others as doctors and nurses in Riyadh. US$7 billion is returned to the Philippines annually as remittances. Reports of ill-treatment of some migrants suggests there is a human cost to interdependency, however.

The benefits and risks of interdependency

Interdependency can strengthen the friendship between states. This may reduce prospects for geopolitical conflict and deliver mutually shared growth and stability instead.

- The first steps towards the EU were taken shortly after the Second World War ended. European state governments believed that greater interdependency would bring centuries of conflict finally to an end.
- Writing in the 1990s, Thomas Friedman argued that economic and political interdependency are linked. In the **golden arches theory** of conflict prevention he asserted that two countries with McDonald's restaurants would never wage war because their economies had become interlinked. While the recent conflict between Russia and Ukraine has weakened Friedman's argument (both countries have McDonald's restaurants), it remains an idea worth exploring.
- States that are home to a large diaspora population often have strong geopolitical ties with the diaspora's country of origin. The enduring friendship between India and the UK is one example. Ex-US presidents Barack Obama and John F Kennedy fostered good diplomatic relations between the Republic of Ireland and the USA while in office (both men being of Irish descent). The arrival of a large Korean diaspora population in the USA has deepened the country's friendship with South Korea.

There are risks associated with interdependency too, however.

- The UK entered recession in 2009 during the GFC (see p. 14). Many building projects were cancelled. The knock-on effect was that many migrants working in construction industries lost their jobs and stopped sending remittances home. As a result, Estonia's economy shrank by 13%. This highlights the challenges that accompany the benefits of interdependency.
- Many citizens view interdependency as a threat to their nation's **sovereignty**. For EU nations, the recent renewal of **nationalism** is linked with a broader debate about 'loss of sovereignty'. A large proportion of citizens in each EU country would like to end the freedom of movement brought by the Schengen Agreement. They believe too much in-migration has been allowed to take place.

Migration policies of host and source countries

Host countries differ greatly in terms of how liberal their international migration rules are. Laws governing economic migration vary over time in line with changes in workforce needs. The UK government adopted a broadly 'open door' approach to international migration in the 1950s and again during the early 2000s. Both decisions related in part to skills and labour shortages arising during those historical periods.

Few countries have laws preventing the out-movement of people because this contravenes the United Nation's **Universal Declaration of Human Rights (UDHR)**: Article 13 guarantees: 'Everyone has the right to leave any country,

Knowledge check 7

Do the migrations shown in Table 8 benefit all the states and people to the same extent? Are these states equally dependent on one another or are the relationships perhaps less balanced than the word 'interdependency' really suggests?

Sovereignty is the freedom of a state to govern itself fully, independent of any foreign power.

Nationalism is a political movement focused on national independence or the abandonment of policies that are viewed by some people as a threat to national sovereignty or national culture.

Exam tip

If an exam question asks you to write about the consequences of a migration movement, check you have read it carefully: there may be a restriction asking you to write only about impacts for the host country, or social impacts, for instance.

including his own, and to return to his country'. As a result, almost all states are, in theory, potential source countries for unlimited out-migration.

■ One notable exception to this rule is North Korea, whose government requires that its citizens obtain an **exit visa** before being allowed to leave.

■ In the past, citizens of the Soviet Union faced similar restrictions on their freedom of movement.

■ In Saudi Arabia and Qatar, some foreign migrants must apply for an exit visa before being allowed to go home.

Rise of anti-immigration movements

In some countries, anti-globalisation political movements are increasingly popular with voters. For many people, the growing rationale for retreating from globalisation is rooted in the valid concern that national cultural identity may be threatened. Migration, in particular, creates political tensions because of differing perceptions of, and viewpoints on, the cultural changes it brings. New political movements in the EU and USA share a common aim, which is to 'regain control' of their borders. Theirs is a nationalist philosophy that is relatively lacking in enthusiasm for multiculturalism and internationally minded politics (or 'internationalism').

■ When a majority of UK voters chose to leave the EU in a referendum on membership in 2016, immigration was the most important issue influencing how people voted. Of London's 8 million residents 30% were born in another country; some Londoners judge the scale and rate of cultural change to have been too great.

■ In many other EU states, nationalist parties now command significant support. In the UK, this takes the form of the UK Independence Party (UKIP), while France's National Front party now has the support of 45% of working-class (or 'blue-collar') voters (but less than 20% of professionals).

■ In France in 2015, staff of the satirical magazine *Charlie Hebdo* were killed by gunmen of Algerian descent. The murderers said their Islamic faith had been mocked. Extreme events such as these are still rare but demonstrate tensions in multicultural Europe that may ultimately threaten the survival of free movement of people.

The USA provides an example of how migration policies can change over time. In the past, it was relatively easy for migrants to enter the country and gain US citizenship. Around 50 million people live in the USA currently who were not born there; over 200 million more are descendants of migrants. Recently, however, the coveted US Green Card has become harder to gain. The issue of illegal migration across the Mexican border is a major policy issue that divides the US public and its politicians alike (Table 9). While in office, President Obama called for work permits to be issued to many of the estimated 8 million unauthorised workers living in the USA. In contrast, during his election campaign, President Trump demanded that a wall be built along the Mexican border.

Table 9 Migration issues that divide popular opinion in the USA

Economic impacts	One view is that migrants are a vital part of the US economy's growth engine. From New York restaurant kitchens to California's vineyards, legal and illegal migrants work long hours for low pay. However, high unemployment in some deindustrialised cities has led to calls for American jobs for American citizens.
National security	The terrorist attacks on the USA in 2001 led to heightened security concerns. Support grew for the anti-immigration rhetoric of the 'Tea Party' movement and, more recently, President Trump.
Demographic impacts	In the USA and other developed countries, youthful migration helps offset the costs of an ageing population. Yet the higher birth rates of some immigrant communities are changing the ethnic population composition of the USA. In 1950, 3 million US citizens were Hispanic. Today, the figure has reached 60 million. This is more than one-fifth of the population.
Cultural change	Migrants change places by influencing food, music and language. Hispanic population growth is affecting the content of US media as programmers and advertisers seek out a larger share of the audience by offering Spanish-language soap operas on channels such as Netflix.

Conflicting views about migration at varying geographic scales

Views about the desirability of migration and cultural change vary from place to place more than many people may realise. Deep schisms in British society were revealed by the UK referendum on EU membership. The country was not united in its desire to quit. Support for 'Brexit' was high among pensioners, rural communities and urban areas in northern England, whereas younger voters, Scots and the cities of London and Cardiff favoured remaining.

In the USA, too, the divisions that have emerged are geographically and sociologically complex. Coastal states such as California and places with a high proportion of Hispanic voters supported Hilary Clinton's broadly 'business as usual' pro-globalisation manifesto in the 2016 presidential election, while some rural interior states and deindustrialised urban areas with the highest proportion of white, poorer and older voters chose Trump to be their leader.

One reason why communities in different places have varying views on migration is because of their own personal experience — or lack of it. Figure 16 provides an interesting insight into how and why attitudes vary in the UK.

- Figure 16 shows public opinion at the level of UK parliamentary constituencies and data on the proportion of people in each constituency who are EU migrants (a large circle means a high proportion).
- The colour of the dots shows people's responses in a survey to the question: 'Do you think that immigration undermines or enriches Britain's cultural life?' Green means more in favour of migration, orange means more against it.
- Researchers found a statistically significant positive correlation between the two variables: British-born people living in cities with a large migrant population, such as London and Cardiff, tend to have more positive feelings (and were more likely to vote to remain in the EU in 2016). British-born people in places with fewer migrants tend to have more negative feelings.
- There are anomalies: Peterborough and Boston, for example, have large migrant populations but few people there believe that migration enriches Britain's cultural life.

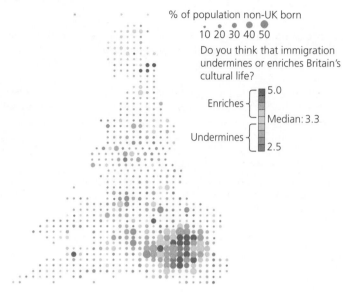

Figure 16 How attitudes towards immigration vary in relation to the size of local migrant populations, 2011

Knowledge check 8

Identify the values shown in Figure 16 for the place where you live. Are these values a surprise or are they what you expected? What factors may help to explain these values?

Self-study task 2

Apply your geographical skills to estimate the **range** of percentages shown in Figure 16.

Summary

- Migration is an important process that contributes to the growing interdependency of places over time.
- Migration can help to reduce inequalities between places; but the backwash process can exacerbate inequalities too, bringing increased economic and political instability.
- As a result, there may be conflicting perspectives on the desirability of the free movement of people.

- Political barriers to global flows of people have both increased and decreased over time in different countries. The migration policies introduced by states reflect both the economic needs of those countries and popular attitudes towards migration held by the electorate.

■ Refugee movements

The causes of forced migration

Refugees are people who have been forced to leave their country. They are defined and protected under international law, and must not be expelled or returned to situations where their life and freedom are at risk. In addition to refugees, many people worldwide have become **internally displaced persons (IDPs)** after fleeing their homes. In 2016, the conflict in Syria, which started in 2011, had generated 5 million refugees and 6 million IDPs; half of those affected were children.

According to United Nations data:

- in 2014, more people were forced to migrate than in any other year since the Second World War. Fourteen million people were driven from their homes by natural disasters and conflict. On average, 24 people were forced to flee their homes each minute, four times more than a decade earlier.

- the global total of displaced people now exceeds 60 million. Of these, around 40 million are internally displaced and 20 million are refugees.
- recent forced movements of people have been caused by wars in Syria, South Sudan, Yemen, Burundi, Ukraine and Central African Republic. Thousands more have fled violence in Central America.

Geopolitical causes in central Africa

Between 1945 and 1970, most previously colonised African countries finally gained their freedom from European superpower rule and became independent sovereign states. In the decades since then, widespread geopolitical instability — especially in central Africa — has led to the widespread forced migration of people. Today, there are more than 2 million refugees scattered across Africa. Large numbers are living in Chad, Kenya and Sudan; significant source countries include Somalia, Rwanda and Angola.

Much of the African continent is afflicted by instability as a result of the way state boundaries were drawn in haste by Europeans hundreds of years ago. In doing so, they gave little or no consideration to the people actually living there (Figure 17). The colonial powers were more concerned with dividing up Africa's raw materials and water resources among themselves. For instance, the boundary between Egypt and

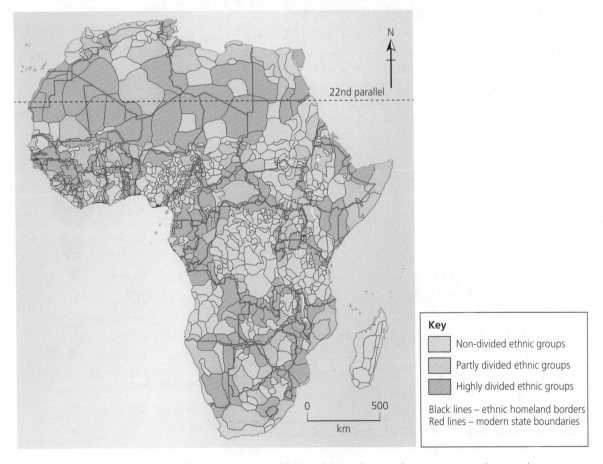

Figure 17 An ethno-linguistic map of Africa, with modern nation states superimposed to show their inherent cultural diversity

Sudan is a straight line drawn for convenience by Great Britain in 1899. It is part of the 22nd parallel north circle of latitude.

- By 1900, many African ethnic groups found themselves living in newly formed nations that in no way represented their own heritage.
- Some long-established ethnic regions were spilt into two or more parts, with each becoming part of a different newly established territory.

This poor approach to state building has played a role in many of the conflicts that followed independence for African countries. For instance, before and since independence in 1962, the Hutu and Tutsi ethnic groups struggled for supremacy in the tiny state of Rwanda. During the civil war of 1994, the Hutu massacred around 800,000 Tutsi. Following this genocide, ethnic Tutsi people in neighbouring Uganda joined with Rwandan Tutsis and fought back. This prompted 2 million Hutu to flee the country.

Self-study task 3

Analyse how the ethnic 'homelands' shown in Figure 17 have been divided by modern state boundary lines. Use examples to support your answer. Why might refugee movements sometimes result from this 'mismatch' between boundaries?

Geopolitical conflict in the Middle East

In recent years, the Middle East has become the world's largest source region for refugees. Internal fighting and interventions by powerful external states, including the USA and Russia, have caused the flight of many millions of people. Figure 18 shows the prominence of Syria and Afghanistan in the list of countries that have generated most refugees on a rolling annual basis since the 1970s. The fluctuations shown for Afghanistan indicate periods when many people fled the state but also years when large numbers returned home.

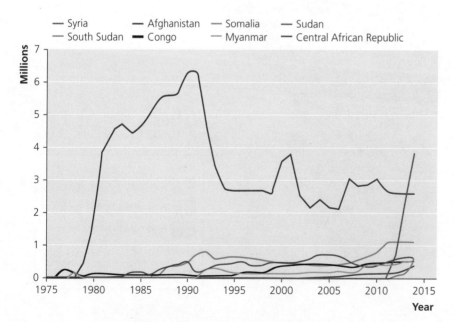

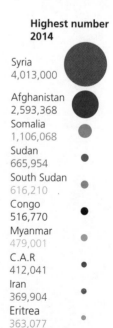

Highest number 2014

Syria 4,013,000
Afghanistan 2,593,368
Somalia 1,106,068
Sudan 665,954
South Sudan 616,210
Congo 516,770
Myanmar 479,001
C.A.R 412,041
Iran 369,904
Eritrea 363,077

Figure 18 The main source countries for refugees, 1975–2014

- Like central Africa, there is an unhappy fit between state borders and the Middle East's ethnic, cultural and religious map. The Sykes–Picot line was drawn by the British and French in 1916. It split apart large Sunni and Shia Muslim communities and led to the creation of several inherently unstable states, including Iraq and Syria. The BBC calls it: 'The map that spawned a century of resentment'.
- The ongoing Israel–Palestine conflict is another major pressure point.
- The stakes have been raised further by ongoing superpower involvement. The USA, Russia, China and some EU countries have a long history of supporting different oil-rich Middle Eastern states and groups. Critics say these superpowers are motivated by their own **energy security** concerns.
- Destabilising terrorist groups have arisen amid the chaos, including al-Qaeda in the Arabian Peninsula and the Taliban in Afghanistan. Daesh (or IS) wages its so-called jihad against all other religions. Its soldiers have pursued a strategy of annihilating minority communities including Christian Assyrians, Kurds, Shabaks, Turkmens and Yazidis. Many people have become refugees.
- The crisis in Syria began when rebel groups demanded the resignation of Syria's ruling President Bashar al-Assad in 2011. The EU and USA initially showed support for some rebels but by 2015 found themselves bombing Daesh in Syria, effectively acting alongside Assad's forces. In 2017, the USA bombed Assad's forces in response to their alleged use of chemical weapons. Meanwhile, Russia and Saudi Arabia have provided funding for rival armies of groups, fuelling the conflict further.

Land grabbing

Refugee movements arise also because of **land grabbing**.

- This is an economic injustice which involves the acquisition of large areas of land in developing countries by domestic industries and MNCs, governments and individuals. In some instances, land is simply seized from vulnerable groups by powerful forces and not paid for.
- **Indigenous groups**, such as subsistence farming communities, may have no actual legal claim to their ancestral land. They sometimes lack the literacy and education needed to defend their rights in a court of law. Around the world there are many instances of unjust land grabs resulting in social displacements and refugee flows. You may be familiar already with the example of Amazonian rainforest tribes losing their land to logging companies.

Drought and climate change

Climate change acts to intensify rural poverty and conflict in some countries. Movers who might previously have been classed as economic migrants become refugees because of an increasingly hostile environment.

- Sudan's semi-arid Darfur region is home to black African farmers and nomadic Arab groups. Between 2003 and 2005, land grabbing and conflict led to the displacement of 2 million people. In this case, competition over land was exacerbated by drought, **desertification** and shrinking water supplies.
- Since 1990, millions of refugees have moved to escape drought in the Horn of Africa; many have moved from Somalia and Ethiopia into neighbouring Kenya.
- Syria's refugee crisis has in part been attributed to desertification by the US Pentagon's security analysts.

Knowledge check 9

How *interlinked* are the different *causal* factors of climate change, drought, land grabs and geopolitical conflict in the examples you have studied?

The consequences of forced migration movements

Social impact on migrants

Many millions of people worldwide are living in camps for IDPs or refugees. In 2016, the countries with the largest numbers of internally displaced people were Colombia (7 million), Syria (7 million) and Iraq (4.5 million). Forced to flee their homes and possessions, IDPs and refugees suffer serious economic and social losses.

- In many camps, adults are unable to work: there are simply no opportunities to make a living.
- Children may cease to be schooled: this has highly damaging long-term impacts for these individuals and their societies. By one estimate, half of all forced migrants are aged under 17 and as many as 90% no longer receive any education or a satisfactory schooling.
- Life in refugee camps can be tough for vulnerable groups including the elderly, the very young and women.
- Human Rights Watch is a **non-government organisation** (**NGO**) which has documented the rape of women and girls living in IDPs camps in Maiduguri, in northeast Nigeria. Camp dwellers also live in constant fear of further attacks from the Boko Haram militia who drove them from their homes originally.
- Many refugees have escaped horrific conditions and continue to suffer trauma as a result. This includes large numbers of people who were forced to fight as child soldiers in Sierra Leone and DRC in the 1990s.

Militia are armed, non-official or informal military forces raised by members of civil society. Militia groups may be characterised as either freedom fighters or terrorists in varying political contexts, or by different observers.

Impacts on neighbouring states

The majority of refugees do not attempt an ambitious long journey to a distant developed country. Instead, they travel no further than the nearest state neighbouring their home country. For families with young children and sick, injured or elderly relatives, it is easy to see why this is the case. Figure 19 shows how the conflict in Syria has put far greater pressure on Turkey, Lebanon and Jordan than on EU states.

Impact on developed countries

A minority of forced migrants gain the opportunity to start a new life after being granted asylum in a developed country (often following a long, dangerous and expensive journey to get there). Catering for refugees can be a major challenge for the host countries and their governments, however.

Since 2006, rising numbers of migrants from north Africa and the Middle East have attempted to reach Europe by crossing the Mediterranean in unsafe fishing boats (Figure 20). By 2016, an estimated 1 million people had attempted the crossing, including many refugees of varying faiths and ethnicities from Syria and poor, war-torn African nations such as Somalia, Eritrea and Ethiopia. More refugees have walked all the way to Europe from Syria, arriving in large numbers at the borders of Hungary and Serbia in 2015 and 2016.

Refugees must apply for asylum (the right to remain) when they arrive in a country. In the UK, a person is only officially designated a refugee when they have their claim for asylum accepted by the government.

Many issues arise for European countries as a result of this mass movement of people.

- EU coastguards have struggled to prevent deaths at sea in the Mediterranean. 800 people died when a boat capsized in rough seas off the Italian coast near Lampedusa in April 2015. By the end of the same year, around 3,700 people had died in similar circumstances. A further 160,000 people were rescued at sea.

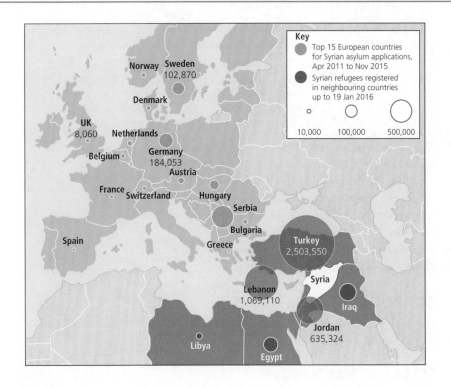

Self-study task 4

Describe the pattern of Syrian refugees shown in Figure 19. Include data in your answer.

Figure 19 The distribution of Syrian refugees in Europe and the Middle East, 2016

Italy
Asylum applications, top five countries (Jan 2009–Jun 15)

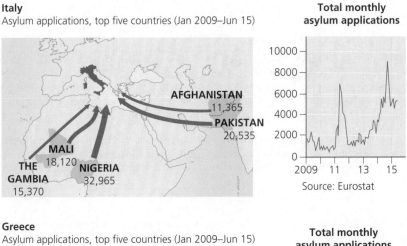

Total monthly asylum applications

Source: Eurostat

Greece
Asylum applications, top five countries (Jan 2009–Jun 15)

Total monthly asylum applications

Source: Eurostat

Figure 20 Source regions for refugees reaching Italy and Greece by boat, 2009–2015

- All EU states — along with most other countries — are obliged to take in refugees, irrespective of whatever economic migration rules exist. This is because they have signed the Universal Declaration of Human Rights (UDHR) which guarantees all genuine refugees the right to seek and enjoy asylum from persecution. The cost of care can be considerable, however. It is estimated to cost £15,000 a year to cater for the needs of one recently arrived refugee who may be recovering from severe physical or psychological harm.
- Although the number of refuges granted asylum in 2015 amounted to less than 0.1% of the EU's population, many European citizens are unhappy with what they view as a high number. It has become an emotive and increasingly divisive issue which divides communities and affects people's voting behaviour in elections. The debate intensified after a suicide bomber in the Paris attacks of December 2015 was identified as a Syrian refugee who had travelled to France via Greece.
- However, in the longer term, many forced migrants will find employment and will contribute to the economy of the states that have granted them refuge.

Managing cross-border refugee flows

Actions to tackle the world's current refugee crises are being taken at different geopolitical scales and have met with varying degrees of success.

Global governance of the rights of refugees

The United Nations uses a variety of methods to protect the human rights of refugees globally. Some of these are outlined in Table 10.

Table 10 How the United Nations offers protection to refugees

The Refugee Convention (1951) and Convention Relating to Stateless Persons (1954)	- The 1951 Refugee Convention is the key legal document that forms the basis of all UN work in support of refugees. Signed by 144 states, it defines the term 'refugee' and outlines the rights of refugees, as well as the legal obligations of states to protect them. The core principle is **non-refoulement**. This means that refugees should not be returned to a country where they face serious threats to their life or freedom. This is now a core rule of international law. - The 1954 Convention Relating to Stateless Persons was designed to ensure that **stateless** people enjoy a minimum set of human rights. It established human rights and minimum standards of treatment for stateless people, including the right to education, employment and housing.
The Office of the United Nations High Commissioner for Refugees (UNHCR)	- UNHCR serves as the 'guardian' of the 1951 Refugee Convention and other associated international laws and agreements. It has a mandate to protect refugees, stateless people and people displaced internally. On a daily basis it helps millions of people worldwide at a cost of around US$5 billion annually. UNHCR works often with the UN's World Health Organization (WHO) to provide camps, shelter, food and medicine to people who have fled conflict. - UNHCR also monitors compliance with the international refugee system. Faced with record displacement from conflict in 2016, UNHCR reinforced the global refugee protection regime by reminding all UN member states of their obligations under international law.
Peacekeeping missions and troops	- UN troops are drawn from the armed forces of many different member states, including the UK, Germany, India and China. Peacekeeping troops sometimes have an important role protecting people in refugee camps from further violence. - Since 1999, up to 30,000 UN peacekeepers have been stationed in DRC: this is the largest-ever deployment of UN troops.

The supporting work of NGOs

The NGO Amnesty International plays an important supporting role by identifying where human rights abuses are taking place. Amnesty lobbies the UN and its Security Council to interview and offer assistance to groups such as the persecuted Rohingya Muslims who fled persecution recently in Burma's Buddhist-majority Rakhine state. Many Rohingya later became effectively stateless people after being stranded at sea in smugglers' boats.

National government policies

Table 11 shows that there are wide variations in the approach of different countries to the challenge of refugee arrivals. Some countries grant asylum to a high proportion of migrants who claim to be refugees; others do not. There is also a wide spectrum of financial assistance provided for those who are classified as genuine refugees.

EU states have struggled to reach a political agreement on where refugees arriving in Greece or Italy should be allowed to settle once their asylum claim has been accepted. Under EU rules, any asylum claims must be processed in the country where refugees arrive. However, neither the Greek nor Italian governments want large numbers of refugees to settle there permanently. Both countries want to see the burden of resettlement shared with other EU members. Consequently, EU states have struggled to reach an agreement on how best to distribute refugees among themselves. In recent years, Germany has taken in many more refugees than any other state: in 2015 alone, the total received was 1.1 million. In contrast, Ireland received just over 2,000 (which is far lower in proportion to its population size).

Table 11 Varying state attitudes towards, and assistance for, refugees seeking asylum. (Sources: UNHCR government statistics/*The Economist* c. 2012–2013)

Country	Applications (decisions made)	Source countries %	Accepted %	Minimum wait before permitted to work	State benefits (single adult, per month)
Germany	173,070 (97,415)	Syria 23 Serbia and Kosovo 14 Eritrea 8	42	3 months	€374 (£325)
Sweden	75,090 (40,015)	Syria 40 Stateless 10 Eritrea 8	77	Immediately, without restrictions	€226 (£197)
Hungary	41,370 (5,445)	Serbia and Kosovo 51 Afghanistan 21 Syria 16	9	9 months, working only in a reception centre	€86 (£75)
UK	31,260 (26,055)	Eritrea 13 Pakistan 11 Syria 8	39	12 months, only jobs where government sees a shortage (medics, engineers, nurses)	€217 (£189)
USA	121,160 (71,765)	Mexico 12 China 11 El Salvador 8	30	6 months, in practice 92% of applicants wait longer for authorisation	€0 (£0)

Self-study task 5

How much variation is shown by the data for host countries in Table 11 and what factors might help explain these variations? You can analyse and suggest reasons for: (a) differences in the source countries for each host country; (b) differences in the number of applications made to each host country; (c) differences in the rules and benefits for refugees seeking asylum in each host country.

The powerlessness of some states to prevent cross-border militia and refugee movements

As we have seen, state building in Africa resulted in fragmented societies. Many ethnic groups, such as the Tutsi and Hutu of central Africa, have a 'transnational' identity. This is an ongoing cause of cross-border movements of people, refugees, militias and armies. Also, there has been little growth in economic development or infrastructure along parts of DRC's northern and eastern borders (in practice, you would not be able to tell where the state boundary lies in many places). As a result of these factors, the governments of DRC, Uganda and Rwanda have been effectively powerless to prevent cross-border flows of people.

■ Armies and militia groups from DRC's nine neighbour states repeatedly entered DRC on the grounds that ethnic groups with whom they claim kinship require support (Figure 21).

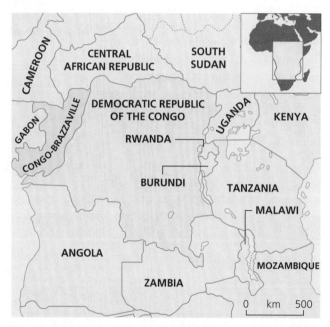

Figure 21 DRC shares a border with nine neighbour countries; refugees flow freely across unmarked state boundaries in forested areas

■ Human greed over DRC's rich natural resources, including diamonds, has attracted unwanted militia groups from other countries. Between the mid-1990s and 2010, millions of people fled their homes in DRC. They were trying to escape attacks by Uganda's invading Lord's Resistance Army (LRA) militia (led by the notorious Joseph Kony). Many fled across DRC's unmarked borders into Central African Republic.

While the situation in DRC has improved in recent years, similar troubles have broken out in northeast Nigeria, where the Boko Haram militia group's campaign of violence and kidnapping has displaced more than 2 million people. Neighbouring countries, including Niger, Chad and Cameroon, have been powerless to stop refugee flows across their own poorly defined borders with Nigeria.

Summary

- There is an important distinction between refugees (people who have been forced to move *between* countries) and IDPs (people forced to move *within* a country).
- Refugees arriving in another country must apply for asylum when they arrive there and prove their life was genuinely at risk in the source country they have left. This forms part of important global governance guidelines established by the United Nations: UN members are expected to follow these rules.
- Refugees are vulnerable people and many are children; their arrival creates a range of short-term challenges for host countries but may bring long-term economic opportunities too (most refugees will work if allowed to).
- Many current refugee flows can be explained by events in the past, including the way African and Middle Eastern states were created under colonial rule.
- Some states are relatively powerless to prevent large-scale movements of refugees and militias across their borders.

Rural–urban migration in developing countries

In terms of the numbers involved, rural–urban migration is the most significant population movement occurring globally. Within a few years, there will be 1 billion rural–urban migrants living in the world's towns and cities. Global urbanisation (the proportion of people living in urban areas) keeps rising (Figure 22).

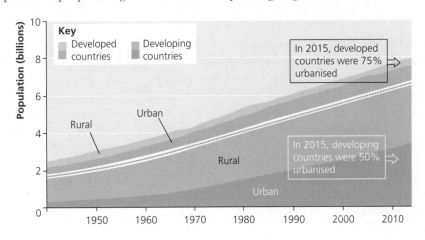

Figure 22 Urbanisation in developed and developing countries, 1945–2015

Push factors in rural areas

Mechanised agriculture, MNCs and new technologies

Global economic systems operate in ways that generate large-scale rural–urban migration flows in many countries. Sometimes this involves the displacement of indigenous people because of land grabbing (see p. 39).

- The main rural push factor is usually poverty, aggravated by land reforms and population growth resulting in not enough jobs being available for those who need them.
- Unable to prove they own their land, subsistence farmers and cattle herders must often relocate to make room for land grabs by MNCs and cash cropping **agribusinesses**. Agricultural modernisation reduces the need for rural labour further, including the introduction of farm machinery by MNCs such as Cargill and Monsanto.
- Rural dwellers are also gaining knowledge of the outside world and its opportunities. The 'shrinking world' technologies we associate with globalisation foster rural–urban migration too. Satellites, television and radio 'switch on' people in remote and impoverished rural areas. As poor individuals in rural Africa and Asia begin to use inexpensive mobile devices, knowledge is being shared. Successful rural–urban migrants communicate useful information and advice to other potential migrants back home in rural areas.
- Transport improvements, such as South America's famous Trans-Amazon Highway, have removed intervening obstacles to migration.

Employment pull factors in urban areas

The main pull factor almost everywhere is employment.

- The **global shift** of employment to Asia, South America and increasingly Africa has created many new work opportunities in cities. In the 1990s and early 2000s, 300 million people left rural regions of China to find work in urban areas.
- MNCs based in developed countries and emerging economies move their own factories and offices to lower-wage locations: this is called **offshoring**.
- MNCs also **outsource** work to foreign companies as part of their global supply chain operations. Outsourcing companies in Chinese cities such as Shenzhen and Dongguan have offered foreign MNCs, including household names such as Bosch, Black and Decker, and Hitachi, the opportunity to have their products made at low prices using a massive pool of low-cost Chinese migrant labour. In recent years, wages have risen in China, however. Some MNCs look increasingly towards Bangladesh and Vietnam instead to find outsourcing partners. In these countries, rural–urban migrants are still willing to work for very low wages. In contrast, manufacturing work in China increasingly involves the production of higher-value goods such as the Apple iPhone.

There may also be **informal** work sector opportunities for new arrivals. Many rural–urban migrants make their living by scavenging landfill sites for recyclable materials; Lagos, Nairobi and Mumbai are examples of where this happens. They sell recyclable plastics and metals to fourth-tier supply-chain companies.

> **Exam tip**
>
> If an exam question asks you to write about the causes of rural–urban migration, you need to provide detailed and specific answers. Saying that the city has 'many jobs and bright lights' may have been a good answer at Key Stage 3. It is not a good A-level answer.

The **global supply chain** is the combination of offshored and outsourced operations on which MNCs rely to produce and procure the commodities and services they sell to customers.

Export processing zones

Asia's three most populated countries — China, India and Indonesia — have all established **special economic zones (SEZs)** where export processing takes place. These serve as strong magnets for rural–urban migration.

- In 1965, India was one of the first countries in Asia to recognise the effectiveness of the export zone model in promoting growth. Today, there are nearly 200 Indian SEZs.
- Coastal SEZs were crucial to China's early economic growth — many of the world's largest MNCs were quick to establish offshore branch plants or build outsourcing relationships with Chinese-owned factories in these low-tax territories. By the 1990s, 50% of China's gross domestic product was being generated in SEZs.
- The low-tax export processing zone in Jakarta is one reason why so much rural–urban migration has been directed towards Indonesia's capital city. It is a popular offshoring location for MNCs such as Gap and Levi's.

A **special economic zone (SEZ)** is an industrial area, often near a coastline, where favourable conditions are created to attract foreign MNCs. These conditions include low tax rates and exemption from tariffs and export duties.

The consequences and management of rural–urban migration

Rural problems and their management

The problems that may result for rural areas are well documented and include the following:

- **Ageing population structure:** India is home to more than 100 million elderly people aged 60 years and above. The majority (around 70%) reside in rural areas. Because of youthful out-migration, relatively few young people remain in some rural districts where large numbers of elderly people need to be cared for.
- **Falling economic productivity:** Youthful out-migration from rural areas in China has left the countryside with an ageing workforce. This threatens future agricultural production and, in turn, the entire country's food security.

One proposed solution to the growing rural–urban imbalance which has been proposed in India and other developing countries is so-called **ruralisation**. This means investing in creating modern, self-sufficient villages that young people are less likely to leave. For this vision to succeed, the state will need to guarantee delivery of essential public services such as roads, drinking water, sanitation, electricity and schools.

Recent step-changes in solar power and mobile internet technology could help make rural areas more attractive places to remain in the future. Kenya's M-Pesa mobile phone service has revolutionised life for local individuals and businesses in rural areas.

- People can transfer money using their phones; the equivalent of around one-half of the country's GDP is now sent through the M-Pesa system annually.
- Fishermen and farmers use mobiles to check market prices before selling produce to buyers. This helps maximise incomes and alleviate rural poverty; more people stay in rural areas as a result.

■ Women in rural areas are able to secure **microloans** from development banks by using their M-Pesa bills as proof that they have a good credit record. This new ability to borrow is playing a vital role in lifting rural families out of poverty. Similar improvements are happening in India and Bangladesh too.

Urban problems and their management

A megacity is home to 10 million people or more. In 1970 there were just three; by 2020 there will be 30 (Figure 23). They grow through a combination of rural–urban migration and natural increase.

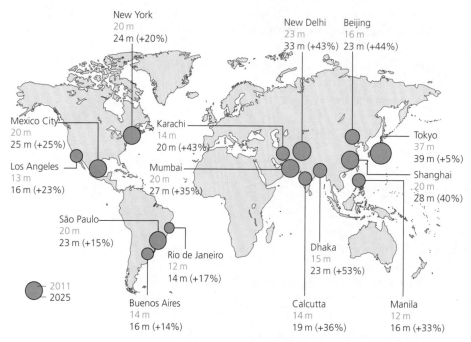

Figure 23 Projected megacity growth over time, 2011–2025 (millions of people)

Megacities in low-income (developing) and middle-income (emerging) countries have grown especially rapidly. São Paulo gains half a million new residents annually from migration. New growth takes place at the fringes of the city where informal (shanty) housing is built by the incomers. **Centripetal migration** brings people to municipal dumps (Lagos), floodplains (São Paulo), cemeteries (Cairo) and steep, dangerous hill slopes (Rio). Over time, informal housing areas may consolidate as expensive and desirable districts. Rio's now-electrified shanty town Rocinha boasts a McDonald's, hair salons and health clinics. In many cities, however, runaway growth of informal housing is a major management challenge (Table 12).

Continued worldwide urban growth is inevitable. As wealth grows in developing and emerging economies, more young rural folk will develop aspirations beyond agriculture. Varying strategies to accommodate urban growth are shown in Table 13.

Table 12 Examples of rapid megacity growth driven by rural–urban migration

Mumbai	Karachi
■ India's Mumbai urban area is now home to 21 million people, having more than doubled in size since 1970. People flock there from impoverished rural states of Uttar Pradesh and Bihar. ■ Urban employment covers a range of economic sectors and skill levels. Big global brands such as Hilton and Starbucks are present in Mumbai. In retail areas, such as Colaba Causeway, large numbers of people work selling goods to the country's rising middle class. ■ Dharavi is a slum housing area in Mumbai. It has a buoyant economy: 5,000 people are employed in Dharavi's plastics recycling industries. However, rising land prices across Mumbai mean there is great pressure to redevelop this and other slum areas.	■ Before Islamabad was founded in 1960, the port city of Karachi was the capital city of Pakistan. ■ Approximately 24 million people lived in Karachi in 2015, making it the most highly populated city in Pakistan and the second most populous megacity in the world (after Tokyo). ■ This colossal megacity is Pakistan's centre of finance, industry and trade. People flock to the city for work from rural areas all over Pakistan, including the Sindh and Punjab provinces. ■ Once living there, rural–urban migrants can find formal or informal employment in a range of industrial sectors including shipping, banking, retailing and manufacturing.

Table 13 Contrasting top-down and bottom-up urban housing strategies

Top-down urban growth strategies	■ China has accommodated rural–urban migration by planning and constructing new housing and cities on an enormous scale. Three major megacity clusters now exist in China's Yangtze River delta (including Shanghai), Pearl River delta (including Shenzhen, which used to be just a fishing village) and the Bohai Sea rim (including Beijing). ■ There are also 60 smaller Chinese cities with populations greater than 1 million. By 2025, according to one estimate, there will be more than 220 Chinese cities with more than 1 million people each. ■ Other countries lack China's wealth and strong top-down leadership, however. Housing shortages are acute in the African megacities of Lagos and Kinshasa, where many people live in slums.
Bottom-up urban community development strategies	■ Poor migrant communities in Lagos must take 'bottom-up' steps to improve their local environment and access to housing by themselves without much state support. As a result, squatter settlements have grown throughout Lagos, and are densely populated because of the shortage of available housing and land. ■ In the case of Makoko, a slum settlement on the edge of Lagos Lagoon, makeshift homes have been built above the water on stilts. People use materials such as tin sheets and wooden planks. They have also reclaimed land from the lagoon using waste materials and sawdust to create new islands for building on. The population of Makoko is estimated at 250,000 people. Most people make a living in the informal economy and by fishing. This goes back to Makoko's origins as a fishing village outside Lagos: as the city has grown, it has been swallowed up in the urban area.

Summary

- The process of internal rural–urban migration in developing countries is linked with the operation of global systems and the actions of global players including MNCs.
- The work of rural–urban migrants is an essential input for the global supply chains on which globalisation depends. In order to maximise profits, MNCs depend on offshored and outsourcing operations in developing world cities where migrant labour can be bought cheaply.

- In the past, much low-paid work was done by rural–urban migrants working in Chinese export zones. Wages in China are now rising, however, and lower-wage Asian countries such as Bangladesh and Vietnam have recently become more important destinations for investment by MNCs.
- Rural–urban migration is expected to continue in many countries well into the twenty-first century; to accommodate predicted urban growth, an effective combination of top-down management and bottom-up community development will be essential.

Global governance of the Earth's oceans

■ Global governance of the Earth's oceans

Supranational institutions for global governance

The meaning of global governance

The term 'governance' suggests broader notions of steering and/or piloting rather than the direct form of control associated with 'government' (Table 14). **Global governance** therefore describes the steering rules, norms, codes and regulations used to regulate human activity at an international level. At this scale, regulation and laws can be tough to enforce, however.

Global agreements and organisations existed in the past. The League of Nations was established after the Great War in 1919, for instance. In the post-war period since the end of the Second World War in 1945 there has been an acceleration towards even greater global governance. At the forefront of this is the United Nations, an umbrella organisation for many global agencies, agreements and treaties.

Table 14 One view of how national governments and global governance differ

	National governments	Global governance
Regulation	Rule of law	Agreements and cooperation
Decision making	Unipolar (centralised)	Multilateral (collective)
Primary goals	Economic growth; national and resource security	World peace; global prosperity and sustainable development

The United Nations, NATO and ocean governance

The **United Nations** was the first post-war **supranational institution** to be established. Over time, its remit has grown to embrace a whole range of areas of governance spanning human rights, health and economics. The **United Nations Educational, Scientific and Cultural Organization** (**UNESCO**) helps protect the environment, including oceans (see p. 82). The UN has also been responsible for the establishment of important **global conventions**:

- The UN Declaration of Human Rights, Human Rights Council and High Commissioner for Refugees (UNHCR) protect human rights and support refugees (see p. 42).

- The 1992 Conference on Environment and Development (the 'Earth Summit') established a plan of action for sustainable development and laid the groundwork for the Kyoto Agreement in 1997 and many subsequent climate change conferences, accords and agreements.

- The UN's global governance framework for Earth's oceans is called the **Convention on the Law of the Sea** (**UNCLOS**). This vast global treaty covers all aspects of marine management, ranging from territorial rights to marine biodiversity. 'Possibly the most significant legal instrument of this century' is how the United Nations Secretary-General described the treaty after its signing in 1982.

The **North Atlantic Treaty Organization** (**NATO**) is another important international agreement, albeit for a smaller group of states. This alliance brings strength in numbers to its 28 member states. Since 1949, the NATO alliance has operated as a mutual defence agreement, meaning that if one member is threatened, all others come to its aid. Several NATO members are major maritime powers, including the UK and USA. NATO takes a special interest in maritime security and plays an active role in tackling piracy and providing assistance to help deal with refugee and migrant crises at sea.

The European Union and the G-groups

Alongside the UN, other smaller groups of states play a significant role in global political, economic and environmental governance. Table 15 evaluates the importance of several of these international organisations, including the EU and various G-groups (Figure 24).

> A **maritime power** is a global superpower or regionally powerful state whose military influence derives in part from the size and strength of its navy.

> **G20 'Group of Twenty'**
> (now 22 members, first formed in 1999)
>
> **G8 'Group of Eight'**
> (now 9 members, formed 1975)
>   France
>   West Germany
>   Italy
>   Japan
>   UK
>   USA
>   Canada (1976)
>   Russia (1997)
>   EU
> Brazil
> China
> India
> Mexico
> South Africa
> Argentina
> Australia
> Indonesia
> Netherlands
> Saudi Arabia
> South Korea
> Spain
> Turkey

Figure 24 Powerful global G-groups

Table 15 Evaluating the role of the EU and G-groups in global governance

Groups	Global role(s)	Evaluation of importance
European Union (EU)	■ The EU has evolved over time from a simple trade bloc into a politically integrated supranational institution with its own currency. ■ The EU is the only group of nations that grants all citizens of member states complete freedom of movement. Most national borders were removed in 1995 when the Schengen Agreement was implemented. ■ The EU has its own strict rules protecting Europe's seas and ocean waters. This is called the Marine Directive.	■ The EU is a highly effective group in many respects because of the high level of political integration its members have achieved. Member states must abide by its economic and environmental laws or face sanctions, including large fines. ■ For instance, ships are forbidden from polluting the territorial waters of any EU country. It is a criminal offence for vessels to discharge oil or other polluting substance. Anyone found breaking this law faces criminal penalties.

→

Table 15 *continued*

Groups	Global role(s)	Evaluation of importance
G7/8 and G20	■ The G8 'Group of Eight' nations includes the USA, Japan, UK, Germany, Italy, France, Canada and Russia (recent conferences without Russia are called G7 meetings). Since 1975, these leading countries with large economies have met periodically to coordinate their response to common economic challenges. ■ In 2011, the G8 acted to stabilise Japan's economy after the devastating tsunami. In 2016, the G7 met to discuss policies capable of stimulating growth in response to the global economic drag caused by China's slowdown.	■ The G7/8 may be losing its importance as a forum for international decision making. This is because several leading economies, including China and India, are not members. ■ A larger group called the G20 was founded in 1999 to include these leading emerging economies in addition to the G7/8 members. ■ However, the large size of the G20, and the differing views of its members, weakens its ability to agree and act on some economic and environmental issues. Critics say it has a poor record on protecting oceans.
G77	■ The G77 group of developing countries in fact has 134 members. ■ This group has lobbied developed countries to do more to tackle climate change.	■ Since forming in 1964, the G77 has functioned as a loose coalition of countries. ■ The diverging interests of its many members has limited its global impact, however.

Ocean governance

Laws and agreements

Various international laws and agreements regulate the use of the Earth's oceans in ways that promote sustainable economic growth and geopolitical stability. The cornerstones of international law pertaining to the oceans are the UN Convention on the Law of the Sea (UNCLOS) and the **exclusive economic zone** (**EEZ**).

The evolution of UNCLOS and the EEZ

Table 16 shows the timeline leading to the global adoption of UNCLOS in 1982.

Table 16 An UNCLOS timeline (Source: www.un.org)

Pre-1939	Prior to the Second World War, the oceans were subject to the 'freedom of-the-seas' principle. States claimed jurisdiction only over a narrow belt of sea (3 miles wide) surrounding their coastlines. The remainder of the seas was proclaimed to be 'free to all and belonging to none'.
1940s–1950s	During this period, there was growing recognition of several threats to oceans, including: ■ pollution and wastes from larger transport ships and oil tankers ■ the toll taken on coastal fish stocks by long-distance fishing fleets ■ new territorial claims over seafloor resources. In 1945, the USA claimed exclusive ownership of its own continental shelf and the oil, gas and minerals found there. This was the first major challenge to the freedom-of-the-seas doctrine. Other nations quickly copied this approach.
1960s–1970s	■ More countries began to claim larger areas of territorial water for themselves. ■ Fish stocks began to show critically endangered signs of depletion. ■ Offshore oil in the North Sea brought the UK, Denmark and Germany into political conflict because of lack of legal clarity over how to share Europe's continental shelf resources.
1973–1982	The United Nations developed UNCLOS: a comprehensive treaty designed to tackle marine pollution, overfishing and competing territorial claims between states, including superpowers.

You will read about several specific UNCLOS provisions later in this book. They include: navigational rights and territorial sea limits; economic jurisdiction and the legal status of resources on the seabed beyond the limits of national jurisdiction; the conservation and management of living marine resources and the marine environment; and the procedure for settlement of disputes between states.

Perhaps the most important contribution of UNCLOS to global governance has been the establishment of the EEZ, which is defined as the area of water extending 200 nautical miles from a state's shoreline (Figure 25). It gives coastal states legal ownership of nearby ocean resources. The coastal state has the right to exploit, develop, manage and conserve all resources — both **biotic** (fish) and **abiotic** (oil, gas or minerals) found in the water, or on the ocean floor, of the EEZ.

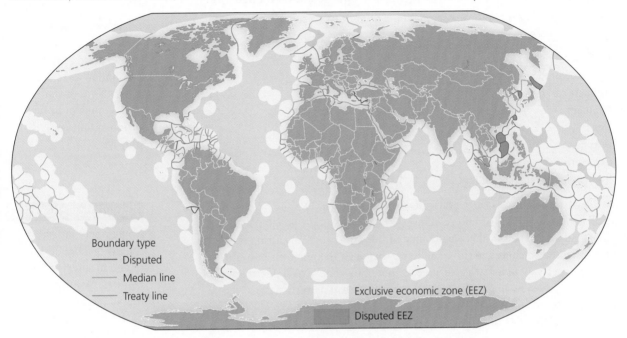

Boundary type
— Disputed
— Median line
— Treaty line

Exclusive economic zone (EEZ)

Disputed EEZ

Figure 25 The EEZ surrounding coastal states and island territories

Sustainability agreements

Important contributions have been made towards sustainable use of the oceans and marine ecosystems by other UN agreements, including the 2015 sustainable development goals (SDGs) and CITES (Table 17).

Table 17 UN actions to help the environment can offer additional protection for threatened ocean species

Sustainable development goals (SDGs)	The UN's 17 sustainable development goals were introduced in 2015. They replace and extend the earlier millennium development goals (MDGs), which were a set of targets agreed in 2000 by world leaders. Both the SDGs and earlier MDGs provide a 'roadmap' for human development by setting out priorities for action. There is a strong emphasis on helping protect the oceans thanks to goal 14: 'Conserve and sustainably use the oceans, seas and marine resources'.
Convention on International Trade in Endangered Species of Wild Fauna and Flora (CITES)	CITES entered into force in 1975. It banned trade in threatened species and their products. Now adopted by 181 countries, it has effectively saved some species (including the Hawaiian nene bird) but not others. Rising wealth in Asia has actually increased trade in some prohibited products, such as the fins of endangered species of shark (see p. 74) and wild turtles. Recently, CITES regulations were strengthened in order to preventing illegal 'finning' of endangered sharks. However, until cultural values shift away from the use of shark fin products, shark numbers are likely to decline further (similar problems exist in relation to growing demand in Asia for rhino horn).

Superpower strategies and security issues

Maritime superpowers

The USA and China are contemporary maritime superpowers. Table 18 compares the size and strength of their navies. Control of the oceans is valued by powerful states as a way of increasing and safeguarding their global **spheres of influence**.

Table 18 Comparing Chinese and US maritime power, 2013 (Source: Military Balance 2013)

	China	USA
Navy personnel	255,000	332,800
Surface combatants (total)	77	112
Aircraft carriers	1	11
Cruisers	–	22
Destroyers	14	62
Frigates	62	17
Submarines	65	72
Patrol and coastal combatants	211+	41

In the past, the UK was the world's greatest naval power. Despite being a relatively small country, by 1920 Britain ruled over 20% of the world's population and 25% of its land area. The Royal Navy dominated the world's oceans during this period, protecting the colonies and the trade routes between them and Britain. In 1914, Britain's navy was about twice as large as that of its closest rival, Germany. The British Empire grew in two distinct phases using naval power:

- **Pre-1850:** Small colonies were conquered on coastal fringes and islands, e.g. New England (now the USA), Jamaica, Accra (Ghana) and Bombay (India), and defended by coastal forts. The forts, and navy, protected trade in raw materials (sugar, coffee, tea) and slaves, and safeguarded the economic interests of private trading companies such as the East India Company.
- **1850–1945:** Coastal colonies extended inland, with the conquest of vast territories. Government institutions with British colonial administrators were set up to rule the colonial population. Complex patterns of ocean trade developed, including the export of UK-manufactured goods to new colonial markets.

Connections have endured between ex-British territories since the Empire was dismantled in the 1960s. A global network of often English-speaking countries belong today to the Commonwealth of Nations.

It is noteworthy too that UK cities with a maritime heritage — such as London, Bristol and Liverpool — are culturally diverse settlements. This is a legacy of the way these places served as hubs for trade and later migration into the UK (epitomised by the passage of the HMS *Windrush* from Jamaica to London in 1948). Knowledge of the UK's past role as a maritime power is essential for gaining an understanding of many aspects of contemporary British life.

Oil transit chokepoints

Maritime trade has always been affected periodically by security issues, including war and piracy. Today, important security issues surround the existence of so-called **oil chokepoints**. The US Energy Information Administration (EIA) defines these as: 'Narrow channels along widely used global sea routes, some so narrow that restrictions

Exam tip

The focus of this part of the course is ocean governance: make sure you stay focused tightly on the topic if a question asks you to write about the importance of different organisations or players.

Knowledge check 10

How important is the ownership of island territories for countries wanting to maximise the area of seafloor they can claim ownership over?

are placed on the size of the vessel that can navigate through them. Chokepoints are a critical part of global energy security because of the high volume of petroleum and other liquids transported through their narrow straits.'

■ About 63% of the world's oil production moves on maritime routes. The Strait of Hormuz and the Strait of Malacca are the world's most important strategic chokepoints by volume of oil transit (Figure 26).

■ Disruptions to these routes can affect oil prices and lengthen shipping journeys.

The Panama Canal was recently made deeper in order to accommodate larger vessels. it was believed this would ease pressure on Middle Eastern chokepoints. The Panama Canal Authority needs to charge up to US$800,000 per large vessel in order to recoup its costs, however. The global shipping industry is struggling currently to remain profitable because of sluggish trade growth, China's slowdown and lower oil prices (see p. 25). As a result, many operators continue to take a longer route via the Suez Canal, which remains cheaper to use than the Panama Canal.

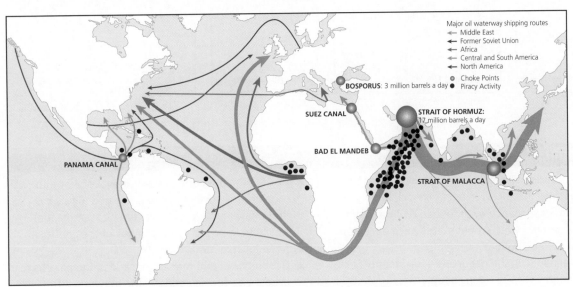

Figure 26 Oil chokepoints, recently recorded piracy activity and proportional arrows showing daily oil shipping

The risk of piracy hotspots

Criminal attacks are a risk to shipping in several hotspots shown in Figure 26. In recent years there has been a marked decrease in piracy attacks along the east African coast (Figure 27) and in the Gulf of Aden. The problem — attributed mainly to poverty and civil war in Somalia — peaked in 2011 when, at one time, 736 hostages and 32 ships were being held for ransom in anchorages off Somali beaches. The same year, the annual cost of piracy — including ransom payments, insurance premiums, the cost of stolen property and various adaptation measures by ship owners — was estimated at around US$10 billion. The Somali pirates were eventually deterred by the interconnected efforts of governments, international organisations and ship-owners:

■ State governments and international institutions including NATO deployed more naval patrols.

■ Ship owners reinforced their vessels with barbed wire, water cannons and armed guards.

■ Ships were rerouted and driven at higher speeds to make it more difficult to board them.

While east African crime has fallen, attacks in southeast Asia — especially along the Indonesian coastline — have risen. Nearly six of every ten sea crimes worldwide happens there. Pirates siphon oil from slow-moving tankers which they capture. By one estimate, pirates stole 16,000 tonnes of oil products valued at US$5 million in 2015. Crews are badly hurt sometimes because these pirates have little interest in ransoming them.

Self-study task 6

Identify the main patterns and trends shown in Figure 27.

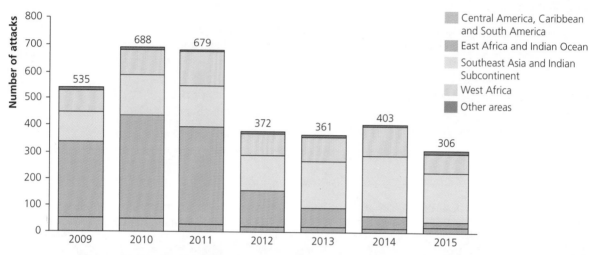

Figure 27 Piracy attacks patterns and trends, 2009–2015

Summary

- Numerous supranational institutions exist. Together, they provide a framework for global governance by establishing important agreements which most states have agreed to respect and uphold.
- Governance of the oceans is achieved primarily through UNCLOS. Many people view UNCLOS as a success story for global governance.
- Both in the past and present, naval power and maritime security have been valued highly by state governments. The risk of piracy adjacent to oil chokepoints is a reminder of the continued need for maritime security.
- Ocean travel and maritime trade in the past have created enduring connections between different countries, places and people.

■ Global flows of shipping and sea cables

Globalisation and trans-ocean commodity flows

Global trade patterns and networks

Trade is the movement of goods and services from producers to consumers. It spans many different sectors of industry. Trade in physical goods (as opposed to services) includes movements of:

- primary industry products (food, energy and raw materials)
- manufactured items (ranging from processed food to electronics)

Overall, world trade is dominated by developed countries and several large emerging economies (EEs) including the **BRIC group** (the four large economies of Brazil, Russia, India and China). The following points provide an overview of global trade patterns, with consideration of both production (origin) and consumption (market).

- The value of world trade and global GDP has risen by around 2% annually since 1945, with the exception of 2008–2009, when the global financial crisis (GFC) led to a brief fall in activity.

- Just ten nations, including China, the USA, Germany and Japan, account for more than half of all global trade.

- The majority of trade originating in developed countries takes place with other developed countries (Figure 28). This is because of the large numbers of affluent consumers and markets found in the world's wealthiest countries.

- Consumer markets have expanded in emerging economies as spending power has grown among their citizens. Middle-class diets are characterised by greater consumption of meat and dairy (higher protein). China's annual meat consumption per capita rose from 4 kg to 52 kg and Brazil's from 28 kg to 82 kg between 1990 and 2010.

- Although its growth is now slowing, China is still the world's number one exporter of goods (valued at US$2 trillion in 2013). Since the early 1980s, China has emerged as the dominant influence on world trade. Indeed, a slowdown in the rate of Chinese growth since 2010 has been responsible for a 'cooling off' of the global economy as a whole. In particular, falling Chinese demand for imports of natural resources and oil has been financially harmful for many African exporters.

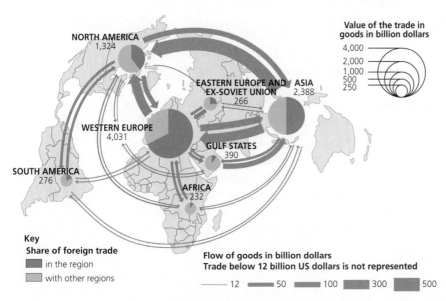

Figure 28 Global trade in goods, 2015

Knowledge check 11

What proportion of the world's intraregional commodity trade flows takes place over oceans? Which are the largest trans-ocean flows, and why?

Self-study task 7

Figure 28 shows the pattern of commodity movements between places. The arrow indicates direction of movement and the width of the arrow is drawn proportional to the volume of movement. Use the information to estimate the value of the five largest trade flows between world regions.

Container shipping trends

More than 600 million individual containers are moved across the oceans each year (Figure 29). Some commentators describe shipping as being the 'backbone' of the global economy ever since the industry's pioneer, Malcolm McLean, loaded the first containers onto one of his company's vessels in Newark, New Jersey, in 1956. Today, everything from chicken drumsticks to patio heaters is transported efficiently across the planet using **intermodal containers**.

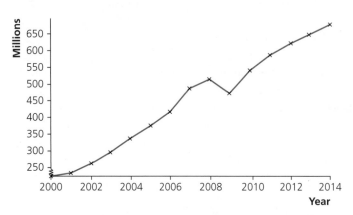

Figure 29 Container movement growth, 2000–2014 (millions of 20-foot units)

> **Intermodal containers** are large-capacity storage units that can be transported long distances using multiple types of transport, such as shipping and rail, without the freight being taken out of the container.

Another important trend is the increase in size of container vessels. Fewer vessels are operating than in the past but they carry more goods.

■ The enormous US$190 million European vessel MSC *Oscar* is 395 m long, 48 m wide and can carry 19,000 twenty-foot equivalent unit (TEU) containers. This compares with the 275 m length of the world's largest vessel in 1988, which carried just over 4,000 containers.
■ The average size of container ships has increased by 90% in the past two decades, and total fleet capacity in 2015 was four times that of 2000.

Recently, the shipping industry has entered a period of crisis because of over-capacity. Shipping companies have built new vessels at a far greater rate than global trade has grown by. Container ships that operate like buses on fixed timetables between China and its major markets have been sailing loaded with too few containers to be profitable. As a result, Hanjin Shipping, the world's seventh-largest operator, filed for bankruptcy in 2016. This mismatch between supply and demand for shipping has arisen because:

■ growth in global trade has dropped (see p. 14)
■ supply chains may be shortening because of more MNCs **reshoring** operations as a result of rising costs and **risks**

The result is that many older vessels may need to be scrapped. **Shipbreaking** is the process of dismantling an obsolete vessel. Because of cheaper labour costs and fewer health and safety regulations, the vast majority of ship breaking takes place in Bangladesh and India.

> **Reshoring** is the shortening of global supply chains. It involves MNCs deciding to produce and procure goods and services locally instead of using distant offshore locations and suppliers. Reasons for reshoring may include avoiding investing in places where a conflict has begun or political pressure from governments (this has happened in the USA).

Managing ocean movements

Regulating shipping flows

UNCLOS guarantees all shipping **the right of innocent passage** through the territorial waters of any state. Movement is allowed 'so long as it is not prejudicial to the peace, good order or security of the coastal state'. Regulation has also been needed in order to safeguard the environment from shipping movements, especially oil tankers. In the past, containerised oil movements brought scores of devastating **transboundary pollution events** to territories flanking shipping lanes. The coastal margins of both France and the UK were severely affected by 119,000 tonnes of oil released from the *Torrey Canyon* supertanker in 1967, after it struck a reef in the English Channel en route from Kuwait to the UK's Milford Haven. This was the first major oil spill to make world headlines:

- 15,000 seabirds were killed.
- 80 km of UK beaches and 120 km of French coastline were contaminated.
- it remains the UK's worst-ever environmental disaster to date.

Figure 30 shows the distribution of the 20 largest spills since then and their proximity to major oil shipping lanes. It is noteworthy that 19 of the largest spills recorded occurred before the year 2000 despite an overall increase in oil trading and vessel size since the mid-1980s. In the 1990s there were 358 spills of 7 tonnes and over, resulting in more than 1 million tonnes of oil lost; whereas in the 2000s there were 181 spills of 7 tonnes and over, resulting in less than 200,000 tonnes of oil lost. In part this is due to successful regulation under UNCLOS.

- Single-hulled oil tankers that were too easily damaged in the past have been taken out of use. The last major single-hulled disaster occurred in 2002 when the *Prestige* supertanker sank off the Galician coast, leading to the largest environmental disaster in Spain's history.
- It is illegal for ships that have recently delivered oil to use seawater to wash out their tanks (flushing of tanks has been a significant cause of oil pollution along major shipping lanes).

Knowledge check 12

How much growth has there been in container movements in recent years? What caused the decline in 2009 which makes the trend shown by Figure 29 uneven?

Knowledge check 13

What examples of risks can you think of which may help explain the new trend of reshoring by MNCs, which results in reduced use of shipping?

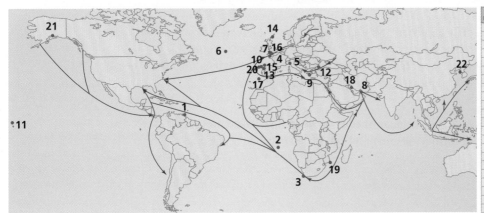

Position	Ship name	Year	Spill size (tonnes)
1	ATLANTIC EMPRESS	1979	287,000
2	ABT SUMMER	1991	260,000
3	CASTILLO DE BELLVER	1983	252,000
4	AMOCO CADIZ	1978	223,000
5	HAVEN	1991	144,000
6	ODYSSEY	1988	132,000
7	TORREY CANYON	1967	119,000
8	SEA STAR	1972	115,000
9	IRENES SERENADE	1980	100,000
10	URQUIOLA	1976	100,000
11	HAWAIIAN PATRIOT	1977	95,000
12	INDEPENDENTA	1979	94,000
13	JAKOB MAERSK	1975	88,000
14	BRAER	1993	85,000
15	AEGEAN SEA	1992	74,000
16	SEA EMPRESS	1996	72,000
17	KHARK 5	1989	70,000
18	NOVA	1985	70,000
19	KATINA P	1992	67,000
20	PRESTIGE	2002	63,000
21	EXXON VALDEZ	1989	37,000
22	HEBEI SPIRIT	2007	11,000

Figure 30 The 22 largest oil spills that have occurred since the *Torrey Canyon* disaster in 1967

Identifying and tackling illegal trans-oceanic flows

Alongside legitimate trade and migration flows, illegal trans-oceanic flows of people, narcotics, counterfeit property, stolen goods and endangered wildlife link societies and places together globally. The United Nations has made repeated calls for states to work together to tackle transnational organised crime flows, many of which use open oceans and territorial waters as their operational space (Figure 31).

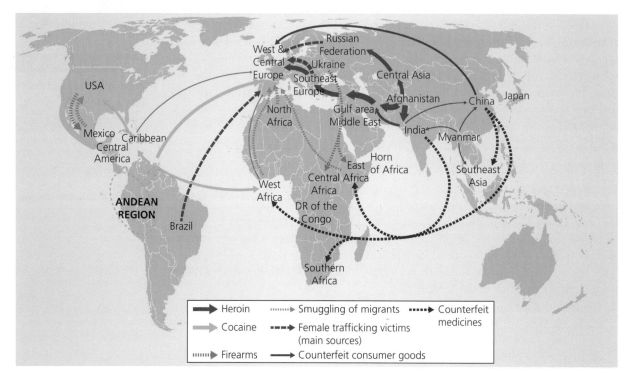

Figure 31 Global transnational organised crime flows (Source: UN)

According to UN Secretary-General Ban Ki-moon:

> Transnational criminal markets crisscross the planet, conveying drugs, arms, trafficked women, toxic waste, stolen natural resources or protected animals' parts. Hundreds of billions of dollars of dirty money flow through the world every year, distorting local economies, corrupting institutions and fuelling conflict. Transnational organized crime markets destroy, bringing disease, violence and misery to exposed regions and vulnerable populations.

The real size of these flows can only be quantified crudely, using police reports and anecdotal evidence. Table 19 includes estimates of the scale and value of selected illegal flows.

Table 19 Examples of illegal trans-oceanic flows and activities

People trafficking	<ul><li>More than 90% of the migrants who cross the Mediterranean illegally use services provided by criminal networks and their associates, according to the security agency Europol.</li><li>It is estimated that in 2015 alone, criminal networks involved in migrant smuggling had a turnover of between €3 billion and €6 billion. Migrant smuggling is a highly profitable business.</li></ul>
Smuggling	<ul><li>Smuggling and unusual shipping activity have increased across the Mediterranean and Atlantic in recent years. Europe has 70,000 km of coastline, much of which is poorly monitored by security agencies.</li><li>This weakness is exploited by organised criminals and terrorist organisations. Illegal drugs, guns and counterfeit goods enter the EU routinely via its coastal margins.</li><li>After the 2001 terror attacks in New York, maritime security standards were strengthened globally with the 2004 International Ship and Port Security Code (ISPS). Introduced by the UN's International Maritime Organization, the ISPS code gives port authorities heightened security powers to monitor shipping and control access for vessels.</li><li>However, much more could to be done to track the movements of shipping in territorial waters. According to the maritime security company Mast: 'If you can get a bunch of AK47 assault rifles into a shipping container somewhere in the world, then you could get them into Europe pretty easily'.</li></ul>
Slavery at sea	<ul><li>There have been allegations of exploitation and slavery in parts of the UK's fishing fleet. The £770 million fishing industry increasingly relies on foreign labour.</li><li>Some foreign workers who do not have permission to live inside the UK are nonetheless allowed legally to work on fishing fleet boats offshore in British waters where they are out of reach of checks by the police or welfare officers. This makes it hard to protect their human rights; some foreign workers from west Africa are being paid so little that their cases may amount to modern slavery.</li></ul>

Information flows

The use of ICT by individuals and societies

The idea of a **shrinking world** and the key role played by ICT in the acceleration of globalisation was introduced on p. 17. The oceans provide the space through which 90% of all internet data passes for a variety of users and purposes (Table 20). If you are watching movies on YouTube or Facebook, there is a good chance that the data have been retrieved from a server in a far-off continent such as North America. In the blink of an eye, information has travelled thousands of kilometres through a **seafloor data cable** to reach your laptop or smartphone. More than 1 million kilometres of flexible undersea cables about the size of garden hoses carry all those emails, searches and tweets. The recent vast expansion of global data flows is shown in Figure 32.

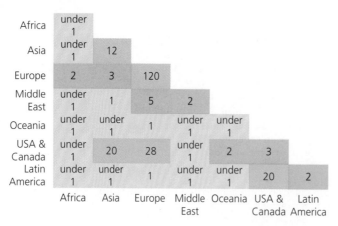

	Africa	Asia	Europe	Middle East	Oceania	USA & Canada	Latin America
Africa	under 1						
Asia	under 1	12					
Europe	2	3	120				
Middle East	under 1	1	5	2			
Oceania	under 1	under 1	1	under 1	under 1		
USA & Canada	under 1	20	28	under 1	2	3	
Latin America	under 1	under 1	1	under 1	under 1	20	2

Figure 32 Global data flows (thousand gigabits per second) between and within world regions, 2015 (Data source: McKinsey)

Self-study task 8

Study Figure 32. Calculate the total volume of data flows linking Europe with other regions. Suggest reasons why the EU is a highly connected world region.

Table 20 Ways in which ICT use supports different facets of globalisation

Economic globalisation	■ ICT has helped MNCs to expand globally by providing connectivity between factories, offices and outsourcing suppliers in different territories. Each time the barcode of a Marks and Spencer food purchase is scanned in a UK store, an automatic adjustment is made to the size of the next order placed with suppliers in distant countries such as Kenya. ■ Media companies can move large data files quickly from animation studios in one country to another, thereby speeding up production time. ■ Economic activity is supported at the personal scale too: self-employed citizens have access to crowdfunding platforms such as Kickstarter to help get their businesses started; they can also sell goods and services globally using markets such as eBay or Amazon.
Social globalisation	■ Facebook, Twitter and Snapchat work by allowing each individual user to function as a hub at the heart of his or her own global or more localised network of friends. ■ Increasing numbers of people gain their education remotely by studying at a virtual school or university, or by enrolling in online MOOCs (massive open online courses). ■ Remote healthcare is being provided in parts of the world where physical infrastructure is lacking.
Cultural globalisation	■ Cultural traits, such as language or music, are adopted, imitated and hybridised faster than ever before. During 2012, South Korean singer Psy clocked up over 1.8 billion online views of *Gangnam Style*, the most-watched music video of all time. ■ Outside of the mainstream, subcultures thrive online too. Small, independent music, comic art and gaming companies can achieve an economy of scale, thanks to a digitally connected global fan base of people sharing the same minority or 'niche' cultural interest. Whether your preference is for folk music from Mali, or a specialist music subculture such as 'grindcore' or 'dubstep', you'll find what you want to hear online.
Political globalisation	■ The work and functioning of multi-governmental organisations (MGOs) is enhanced by the ease with which information and publications can be disseminated. Websites for MGOs such as the EU, UN and World Bank contain a wealth of resources that aim to educate a global audience about issues ranging from climate change to international war crimes. ■ Social networks are used to raise awareness about political issues and to fight for change on a global scale. Environmental charities such as Greenpeace spread their messages online.

Seafloor cable data network growth and management

For more than 150 years, seafloor data cables have helped different continents to communicate in real time.

- The first telegraph cables capable of sending Morse code messages were laid on the Atlantic seafloor in the mid-1800s (they feature in French writer Jules Verne's 1870 novel *20,000 Leagues Under the Sea*).
- The next significant step forward came in the 1950s when coaxial cable, capable of carrying telephone conversations, became the standard.
- In the 1990s, analogue cables were replaced by fibre optic cables capable of carrying huge amounts of digital data in the form of light.
- Today, 99% of all intercontinental data traffic — including phone calls, texts and emails — is transmitted via seafloor cables (Figure 33). Because of cloud-based technologies and greater use of on-demand streamed media by the world's growing middle class of consumers, demand for global bandwidth is growing at up to 40% per year.

Exam tip

Table 20 shows a good way to structure an answer to the following question: 'Explain how information flows have helped to accelerate globalisation.'

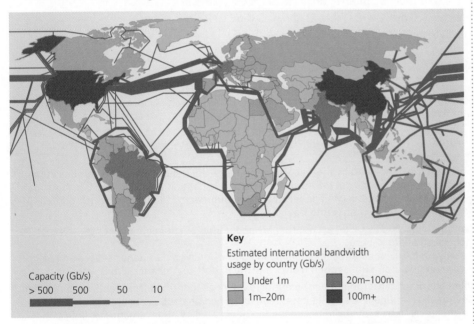

Figure 33 The uneven global distribution of undersea data cables (2012)

Cables used to be laid and controlled by governments; today, MNCs own much of the infrastructure.

- Microsoft and Facebook are jointly building Marea, a 6,600 km, highest-capacity-ever Atlantic cable linking the USA, Europe, Africa, the Middle East and Asia.
- Google helped fund the US$300 million FASTER cable project linking the USA, Japan and Taiwan.
- Vodafone recently laid a new cable from Bengal to southeast Asia, south Asia and the Middle East.
- BT, AT&T and Telefónica all own large seafloor cable networks.

Global governance measures have been in place for more than 100 years that offer protection to seafloor cable networks, including international conventions that have been in place since the 1880s (for telegraph and telephone cables). The first international agreement — the **Convention for the Protection of Submarine Cables** — was signed by around 20 European, North American and South American states in 1884. Today, UNCLOS provides and expands the protections originally accorded to telegraph cables to all international fibre optic cables:

- Data cables are recognised as critically important twenty-first century infrastructure and deserving of special protection; states can establish no-fishing and no-anchoring zones around important cables.
- All states enjoy the freedom to lay and maintain submarine cables in the EEZ and on the continental shelves of other states.
- In the future, restrictions may be introduced on laying cables across vulnerable seafloor ecosystems; as yet, though, no such limitations exist.

Risks to global connectivity

With so much vital information passing through potentially vulnerable networks, it is perhaps surprising that damage and service interruptions are not more common. Thankfully, however, accidents are relatively rare. Table 21 shows several well-documented hazards affecting data networks.

Table 21 Physical and human hazards affecting seafloor data cables

Tectonic hazards and landslides	■ Telecoms network builders must overcome enormous challenges set by physical geography. Modern seabed mapping systems use sonar and high-definition seismic profilers to **mitigate** these risks by avoiding especially hazardous zones in tectonically active areas along the mid-Atlantic ridge. ■ In 2006, a submarine earthquake and landslide destroyed Taiwan's telecom link with the Philippines.
Tsunamis and cyclones	■ Following the tsunami generated by the Andaman–Sumatra earthquake on 26 December 2004, land-based telecommunications networks were damaged in coastal Malaysia and South Africa ■ In 1982, the passage of Hurricane Iwa triggered several submarine landslides, which damaged six seafloor telephone cables connecting Hawaii.
Anchors and trawling	■ The most common hazards by far — accounting for about 60% of cut cable incidents — are dropped anchors and fishing nets ■ Asia temporarily lost 75% of its internet capacity in 2008 when a ship's anchor severed a major internet artery running along the seabed from Palermo in Italy to Alexandria in Egypt.
Fish and shark attacks	■ Fish, including sharks, have a long history of biting cables, as identified from teeth embedded in cable sheathings. Barracuda and sharks have been identified as causes of cable failure. ■ Fish may be encouraged by a 'strumming' sensation generated by electromagnetic fields.
Sabotage	■ Cable sabotage was common during both World Wars.

Summary

- Covering two-thirds of the Earth's surface, the oceans provide a vast space within which two global flows of core importance for globalisation operate: these are commodity flows (via container shipping) and data flows (via seafloor cables networks).
- The patterns and trends associated with these movements and flows are complex. They have accelerated over time but not in a uniform way. Some parts of the world remain poorly connected.
- Numerous human and physical risks threaten global flows of shipping and data; risk mitigation can be achieved by international actions and the strategies of governments and businesses.

◼ Sovereignty of ocean resources

Natural resource distribution and availability

The ocean floor is a source of **abiotic** resources for those countries equipped with the technology needed to exploit them. These natural resources include **fossil fuels** (conventional and unconventional oil and gas) and **minerals**.

Mineral resources

The **coastal waters** of some countries are rich in **placer deposits** which can be recovered relatively easily and cheaply. Placer deposits originate on land and have been transported by rivers into estuaries and nearby coastal waters. For instance:

- diamonds off the southern and western coasts of South Africa
- deposits of tin, titanium and gold found along the shores of Alaska and some South American states

Coastal waters also provide societies with salt, sand and gravel.

Deep sea waters — lying beyond the national jurisdiction of individual states — contain vast amounts of unrecovered mineral resources, including manganese nodules (usually located at depths below 4 km), cobalt crusts (formed along the flanks of undersea mountain ranges) and sulfide muds (occurring along plate boundaries). Table 22 provides details of these important deep seabed resources whose distribution patterns can be linked with tectonic margins and features.

Table 22 Ocean floor abiotic resources

Iron, copper, zinc and gold	These minerals are present in sulfur-rich mud and ores found near ocean-floor **black smokers** at submarine plate boundaries. Along the East Pacific Rise and the Mid-Atlantic Ridge, black smokers produce iron-rich sulfides. Copper, zinc and gold can be found in the southwest Pacific Ocean.
Manganese nodules	Manganese nodules are dense lumps of manganese, iron, silicates and hydroxides, ranging from golf ball to tennis ball size. They grow by just 2 mm every 1 million years as a result of chemical reactions occurring in seawater. In the eastern Pacific Ocean, manganese nodules cover an area of sea floor the size of Europe. Their concentration here is linked with hydrothermal activity at the East Pacific Rise.
Cobalt crusts	Cobalt crusts form at depths of around 1–3 km on the flanks of submarine volcanoes in tectonically active regions such as the South Pacific. Cobalt is found on land in only a few countries. This means that oceanic cobalt is potentially valuable if it can be recovered.

Commercial interest in deep seabed mineral resources has varied over time in accordance with fluctuations in commodities markets and new discoveries of easier-to-access onshore deposits.

- Currently, it is rarely profitable to retrieve and utilise deep sea mineral resources because of the difficulties involved. It is an expensive and not always successful process. The UNCLOS website likens deep seabed mining to: 'Standing atop a New York City skyscraper on a windy day, trying to suck up marbles off the street below with a vacuum cleaner attached to a long hose'.
- Between 2014 and 2016 prices of many minerals fell because of shrinking demand from China, whose economic growth has slowed (see p. 25). This has reduced further the profitability of ocean floor mineral recovery.

Fossil fuels

Ocean floors are important for the recovery of fossil fuels. Important **oil and gas reserves** are found in many shallow and deeper offshore locations (Figure 34). The greatest costs and risks are associated with deeper offshore reserves. The oil spill in the Gulf of Mexico in 2010 demonstrated this: millions of barrels escaped after the oil rig Deepwater Horizon exploded. States that nevertheless permit MNCs to drill for oil in deep water close to their shoreline include Australia, Greenland, Norway, Canada, Libya, China, Brazil and Angola.

New offshore oil reserves are sometimes found using geological theory grounded in an understanding of plate tectonics. According to the Atlantic mirror theory, geological conditions along the coast of Brazil match those at the west African coast. Present-day oil fields off the coasts of Angola and Brazil were originally part of one enormous oil reservoir which split in half when the two continents drifted apart 100 million years ago. When large offshore oil discoveries were made near Brazil in 2006 (in the prolific Santos and Campos Basins), geologists working for BP, Total and Statoil rushed to explore Angola's offshore waters. Several major new oil fields were found there.

Oil and gas reserves are technically and economically recoverable oil and gas fields. Oil and gas in areas of very deep water would not be designated as reserves because of the likelihood that it will never become technically possible to recover them at an acceptable cost.

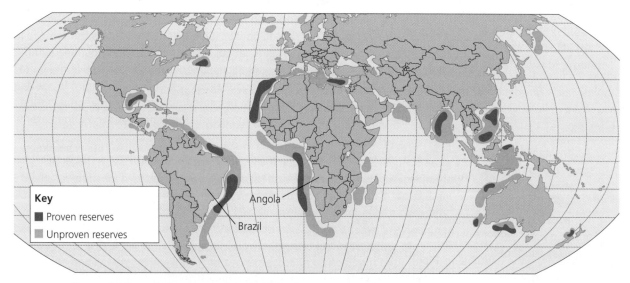

Figure 34 Potentially recoverable fossil fuel reserves are located in many offshore areas

Geopolitical tension and conflict

Coastal countries may claim **sovereign rights** over an exclusive economic zone (EEZ) extending to a **territorial limit** 200 nautical miles from their shoreline (Figure 35). In practice, EEZ ownership claims may still be contested.

- EEZs overlap sometimes when two states lie close to one another. The Sea of Japan (bordered by Japan, South Korea and China) is a case in point (see p. 68).
- Some states own overseas territories and may claim an EEZ around these. The UK has established an EEZ around the Falkland Islands close to the coast of Argentina. The war fought by the two countries in 1982 related in part to contested maritime jurisdiction.

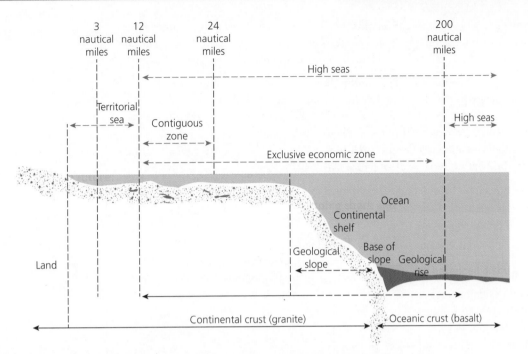

Figure 35 The EEZ (note the continental shelf does not extend beyond the EEZ in the example shown here)

Conflicting claims may arise over ocean floor beyond the EEZ limit. Since 1994, the International Seabed Authority (ISA) has helped map the legal right of states to use ocean floor resources beyond their own territorial waters. Some countries with a large **continental shelf** — including Argentina, Australia, Canada, India, Madagascar, Mexico, Sri Lanka and France — have argued that it should be recognised as an extension of their land territory. The ISA has allowed these countries the possibility of establishing a boundary going out to 350 miles from their shores or further, depending on certain geological criteria, such as the thickness of sedimentary deposits. These additional areas allow for mineral exploitation; they do not alter the legal status of the waters above, which remain High Seas.

Superpower tensions

The topic of global and regional superpowers seeking to safeguard their maritime sphere of influence was explored previously (p. 54). Two important territorial sources of superpower tension in recent years are the islands of the South China Sea and the Arctic Ocean.

The South China Sea

China has repeatedly made territorial claims on parts of the South China Sea that are contested by other states, including the Philippines. Figure 36 shows how China claims jurisdiction over the South China Sea (within the red dashed line) despite the proximity of other states to this area of water.

China's argument is based on its ownership claim of several island groups and their surrounding EEZs, but this policy is contested by China's neighbours. First, there are

A **continental shelf** is a seabed that sometimes extends beyond the limits of the EEZ to the outer edge of a continental tectonic plate margin.

Knowledge check 14

How does the study of plate tectonics at GCSE or A-level help us to understand why some states are allowed ownership of ocean floor resources up to 350 miles from their coastline?

competing claims on some of the islands. Second, some of the islands are little more than rocks, which China has enlarged artificially. For instance, in 2014 China began constructing an airport on reclaimed land on Fiery Cross Reef (in the Spratly Islands).

The South China Sea is strategically and militarily important to China: it is an important **energy pathway**. Chinese **energy security** depends on the passage of oil tankers through these waters. The EEZs around these islands may harbour oil and gas reserves too. As a result, China is unlikely to abandon its 'Nine-dash Line' policy any time soon (so-called in reference to Chinese maps delimiting the area of claimed control in the South China Sea using nine dashes).

China's actions have heightened regional tension in recent years.

- China has begun to question the right of US ships and aircraft to sail and fly in the disputed areas.
- In 2016, a United Nations tribunal in The Hague found that none of China's man-made islands were substantial enough to command 200-nautical-mile EEZs. The tribunal said China had illegally infringed the Philippines's sovereign rights to fish and develop energy resources in its own coastal EEZ. China's response was to brand the UN arbitration as 'unlawful'.
- The Philippines has adopted a policy of blowing up Chinese fishing vessels intercepted in contested parts of the South China Sea.

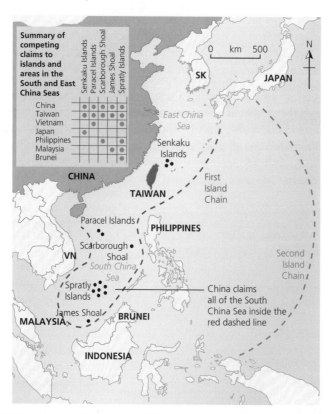

Figure 36 Tensions in the South and East China Seas (Source: Cameron Dunn)

Arctic Ocean issues

Tensions have arisen in recent years over the governance of Arctic Ocean resources. Competing claims have been made over ocean waters and resources by rival superpowers, nations and groups of indigenous people.

The Arctic Ocean is thought to hold about 90 billion barrels of oil. Ice cover is thinning on account of a warming climate and some experts predict the region will be entirely ice-free by 2050. Temperatures 20°C higher than usual in late 2016 suggest it may occur even sooner. Figure 37 shows competing national claims to the Arctic Ocean seabed. Will increased accessibility trigger a 'black gold rush' to recover oil from an ice-free Arctic, resulting in geopolitical tension or conflict?

- One dispute centres around whether an area of ocean bed known as the Lomonosov Ridge is an extension of Russia's continental shelf — and therefore legally deemed an extension of its land territory — or not.
- In 2007, the Russians used a submarine to place their flag on the seabed at the North Pole. This was widely viewed as an aggressive geopolitical action.
- Before leaving office in 2017, however, US President Obama declared a huge area of Arctic waters as 'indefinitely' off limits to oil and gas exploration as part of a joint move with Canada. This could reduce tensions in the region. However, the decision could be reversed by a future administration.

Economics may be a more important factor than politics in determining the Arctic's long-term future, though. The enthusiasm of oil and gas MNCs for Arctic exploration vanished when crude oil prices crashed in 2014. Thereafter, the world price of oil remained lower than the potential cost of extracting offshore Arctic oil. Royal Dutch Shell abandoned offshore drilling in Alaska in 2015 despite having already spent US$7 billion in the Chukchi and Beaufort seas. Citing the high cost and risk of

Knowledge check 15

Why do geopolitical tensions and conflicts sometimes develop over territorial claims in the oceans?

Exam tip

Information about geopolitical case studies can date quickly. If you are reading this book in 2020 or later, it would be a good idea to research what has happened since the book was written. Examiners are impressed by contemporary answers.

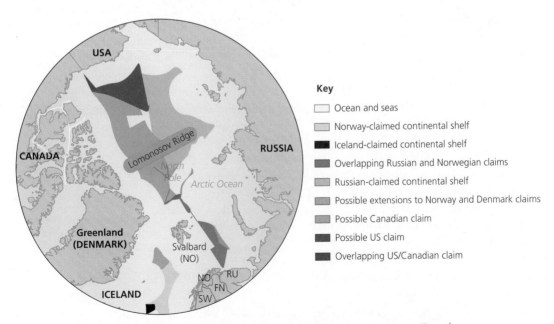

Figure 37 Territorial claims in the Arctic (Source: Cameron Dunn)

Key
- ☐ Ocean and seas
- ☐ Norway-claimed continental shelf
- ■ Iceland-claimed continental shelf
- ☐ Overlapping Russian and Norwegian claims
- ☐ Russian-claimed continental shelf
- ☐ Possible extensions to Norway and Denmark claims
- ☐ Possible Canadian claim
- ■ Possible US claim
- ■ Overlapping US/Canadian claim

working in Arctic waters and the low market value of oil, Shell has ceased operations in the Arctic 'for the foreseeable future'. Even if Obama's ban is overturned in the future, large-scale Arctic oil exploitation is unlikely while energy remains cheap.

Landlocked countries and societies

The relative isolation of landlocked countries can play a role in their economic health, because of the greater difficulties involved in trading without a coastline. Further injustice arises from the unequal access to ocean resources among the world's states and nations. In contrast to their maritime neighbours, the 45 countries that lack a coastline are not entitled to any benefit from the fossil fuel and placer mineral deposits found in the coastal waters of the continents they are part of. It is easy to see why wars have been fought throughout history over the ownership of islands and coastal regions.

Countries without coastlines

With a few exceptions, the world's landlocked countries are relatively poor and have lower levels of trade than maritime countries (those with a coastline).

- Of the 15 lowest-ranking countries in the human development index, eight have no coastline. All are in Africa and their per capita GDP is 40% lower than that of their maritime neighbours.
- Their most obvious handicap is in moving goods to and from ports.
- Some people think landlocked countries have not enjoyed the historic benefits of global flows of migration, ideas and new cultural ideas. Global flows that brought innovation to maritime countries may have largely bypassed landlocked ones.

Some landlocked countries have developed economically and become important global hubs, however. Being landlocked is *not necessarily* a barrier to global interactions. Trade agreements could be more important than location in helping a country to prosper. Landlocked Switzerland is a major global finance centre and the headquarters of many MNCs, including UBS and Credit Suisse. Botswana is a middle-income landlocked country which exports diamonds using global air networks.

Gaining ocean access and trade

Under international law, countries with no sea coast have a right of access to and from the ocean via **transit states** for the purpose of enjoying 'the freedom of the High Seas'. In reality, this freedom can be difficult and expensive to uphold.

- Bolivia lost its Pacific coastline to Chile in a nineteenth-century war.
- Chile promised afterwards to allow 'the fullest and freest' commercial transit. To this day, most of Bolivia's imports and exports pass by lorry through Chile. Unfortunately, this passage can become an obstacle course of delays, including inspections and poorly maintained roads. A strike by Chilean customs officials in 2013 caused a queue of lorries 20 km long in Bolivia.
- By one calculation, Bolivia's GDP would be one-fifth higher if it still had direct access to the sea. It is now the poorest country in South America and blames this state of affairs on its landlocked condition.
- Bolivia has applied to the International Court of Justice (ICJ) in The Hague to be given 'sovereign access to the sea' — in other words, it wants a stretch of its old Pacific Ocean coastline to be returned.

> **Exam tip**
>
> Think about scale here. Not all places in a maritime country benefit from its coastline. Maritime states contain isolated rural areas whose populations are relatively unconnected with the rest of the world either by choice or constraint (such as the Amish community in the USA, or Amazonian tribes in Brazil).

Injustices for indigenous people in coastal areas

Many of the world's coastal margins are home to indigenous populations, some of whom are almost entirely dependent on fishing for their livelihood. Some indigenous communities claim to have been treated unjustly in relation to ocean resource use in the coastal waters they rely on. Others have taken action to protect their rights.

- Along Alaska's 14,000 km of coastline, many Native American communities earn a living from salmon, crab and whitefish fisheries. They also remain dependent on fish for a range of 'non-commercial, customary and traditional uses'. Fish provide Inuit populations with food, oil (for fuel) and bones (used to help make clothing and tools).
- When a new gold-mining development in Alaska's Bristol Bay threatened the area's sockeye salmon ecosystem with water pollution, Native American communities ran an effective 'no dirty gold' campaign.
- To date, they have succeeded in stalling the mine's construction. Many large MNCs, including Tiffany's, have pledged to boycott Pebble Mine gold should the project ever be completed.

Summary

- Over time, the global community has developed a shared set of rules in relation to how ocean and seabed resources are used, and by whom. There is wide agreement on the principle that maritime countries have sovereign rights over the EEZ and, in some cases, a wider continental shelf.
- Deep seabed mining and fossil fuel extraction is challenging and expensive; most economic activity occurs in shallower offshore waters within EEZs.
- Despite the many rules and laws that have been developed, geopolitical tensions remain common. Islands often become contested territories. Physical changes in the Arctic Ocean are creating a new governance challenge because previously inaccessible resources are becoming accessible; and who owns them is unclear.
- Injustices arise from the unequal access to ocean resources: landlocked countries may be poorer because of their lack of an EEZ and coastline.

■ Managing marine environments

The global commons

The four global commons

Large parts of Earth's oceans are a shared space used by all states, also known as a **global commons**. More than 60% of the oceans are designated as 'high seas' by UNCLOS. This makes them **areas beyond national jurisdiction** (**ABNJ**).

It is in the best long-term interest of individual countries to collaborate on making sure that sustainable use of the oceans, and oceanic resources, is achieved over time. However, for the governance of this global commons to be successful, there needs to be some degree of harmonisation between the **norms** and laws of the community of states that use the oceans in varying ways, which includes fishing, mineral and fossil fuel recovery, transport and communications, renewable energy and waste disposal.

International law identifies four global commons: the oceans, the atmosphere, Antarctica and outer space.

Global commons are global resources so large in scale that they lie outside the political reach of any one state.

Norms are widely shared or accepted ways of behaving that most people or governments would agree with.

Changing perspectives on managing the oceans as a global commons

The famous phrase **tragedy of the commons** can be traced back to the writing of Adam Smith in the 1700s. Originally used to describe the unsustainable use of land at a local scale, it refers to what happens when individuals overexploit a shared or common resource — selfishly leading to the long-term damage or utter loss of that land (due to soil erosion, for instance). As a result, everyone loses out. In the 1960s, the ecologist Garrett Hardin argued that the same could happen at a global scale: as world population grows, unsustainable use of shared environments and ecosystems may result in overexploitation and permanent ruin.

Ocean fisheries and marine species are at particular risk of overexploitation. Without controls, too many fish and ocean mammals can be caught before they have had time to breed and reproduce the next generation. This is due to improvements in technology including:

- increased **longlining** (some ships lay 150 km lengths of baited hooks on the seabed)
- the use of **sonar** to detect shoals of fish that might otherwise have been missed
- huge **factory** ships with freezers, which allow ships to stay out at sea longer

To prevent the tragedy of the commons, regulation is required. There are two contrasting approaches that can be taken in order for oceans and their ecosystems to be protected (the same choice also applies to the management of terrestrial environments and ecosystems).

1 The **conservation** management approach allows the efficient, non-wasteful and sustainable use of natural resources. Marine conservation allows continued commercial use of oceans but with limits applied. **Quotas** may be attached to cod fishing for instance. This ensures that sufficient fish are left in the sea to reproduce the next generation. Pollution may need to be banned if it poses a threat to marine life.

2 The **preservation** management approach views nature (including the oceans) as something best left apart from human commerce entirely. This is a 'keep off the grass' philosophy that may involve an outright ban on commercial activity or catching of some species. The attempt to ban whale hunting completely is a preservation strategy. Unfortunately, preservationist attitudes may not offer a sustainable future to communities who rely on exploiting oceans and their ecosystems in order to survive.

Changing attitudes towards the oceans

Both conservation and preservation approaches have their own merits in different contexts. They share a philosophy of environmental stewardship, which sees humans as 'caretakers' of nature. Over time, this worldview has become an increasingly widespread social norm that is now shared by many individuals and societies. Changing attitudes towards whaling provides an example of this (Figure 38).

Many species of whale were hunted almost to extinction during the twentieth century, bringing a public outcry in many countries. Some whale populations have subsequently been helped to recover by global agreements. Since 1946, the whale industry has been regulated by an inter-governmental body called the International Whaling Commission (IWC). In 1986 the IWC issued an indefinite ban on commercial whale hunting. Also, the United Nations Convention on the Law of the Sea (UNCLOS) requires that the 168 nations who have signed it must follow IWC guidelines.

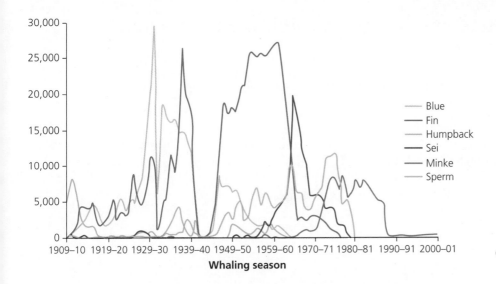

Figure 38 Twentieth-century whale catches in the Antarctic Ocean

The IWC has not been entirely successful in its efforts. Continued whale harvesting is still permitted by some 'indigenous' (traditional) societies. Several countries and territories continue to allow hunting, albeit at lower levels than in the past, including Japan (which has a long history of defying international whaling laws in the interest of what its government calls 'scientific research'), Norway and the Faroe Islands (where around 1,000 pilot whales are massacred annually in line with local tradition). In the majority of countries, though, whale hunting is no longer viewed as acceptable.

The gradual recovery of some whale populations is in line with the predictions of the environmental Kuznets curve (Figure 39). This model suggests that many countries — and by extension the global community — have become more aware of their environmental impact and are acting to reduce their impact over time. However, while actions to protect whales may have had some success, we have a long way to go before the problems of overfishing (p. 75) and marine pollution (pp. 78–79) are tackled satisfactorily.

Self-study task 9

Study Figure 39. To what extent does this model offer hope that Earth's oceans may yet avoid 'the tragedy of the commons'?

Knowledge check 16

What do the examples of continued whaling by Japan, Norway and the Faroe Islands tell us about the limits of global governance? To what extent is the global community of nations in agreement about this or other important issues?

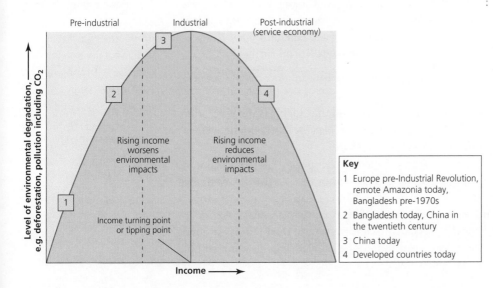

Figure 39 The environmental Kuznets curve

Affluent societies and the overexploitation of marine ecosystems

Population growth, rising affluence and natural resources

Mainly as a result of global trade, prosperity has risen in emerging economies such as India, China, Brazil and the 'MINT' group consisting of Mexico, Indonesia, Nigeria and Turkey. Almost 1 billion people in Africa, Latin America and Asia have attained 'middle-class' status in the last 30 years; 2 billion more are on the cusp of it (Figure 40). Inevitably, this puts increased pressure on ocean resources. Can Earth cope with the growth of consumer societies?

A **consumer society** is a society made up of people who aspire to consume luxury goods and services.

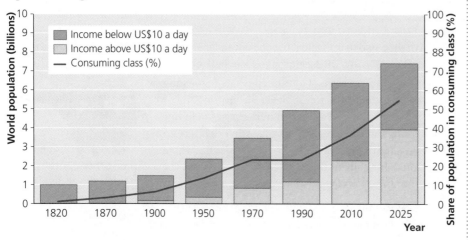

Figure 40 Actual and projected growth of the 'consuming class' or 'new global middle class', 1820–2025

The average US citizen has an ecological footprint 20 times larger than a subsistence farmer in sub-Saharan Africa. In other words, the same area of land that supports ten US citizens with high-impact lifestyles supports 200 low-impact lifestyles. Increasingly, this 'US lifestyle' is an aspiration for people in developing and emerging economies. Inevitably this increases the size of humanity's **ecological footprint** further.

For instance, rising affluence and aspirations in China result in more people adopting a diet that is richer in meat and fish.

- On land, this brings rising water insecurity (farmland that supports ten vegetarian diets will only meet the needs of a single meat-eater, because of the inefficient energy transfer between trophic levels in a food chain).
- In the oceans, the result is a growing list of endangered species, including salmon, tuna and sea turtles. The shark is an unfortunate 'poster boy' for the unsustainable rising trend in consumption that correlates with China's leap out of poverty. In Chinese culture, shark fin soup is a traditional, if expensive, catering choice at weddings. As more marrying couples can afford to serve it to their guests, so shark numbers dwindle further as a result of 'finning'. The problem can be summed up succinctly as: 'new money, old values'.

An **ecological footprint** is a crude measurement of the area of land or water required to provide a person (or society) with the energy, food and resources needed to live, and to also absorb waste. For someone in the UK, it is about the size of six football pitches (the global average is one-third of this).

Another rising pressure linked with global economic development is **invasive species**. Growth of global trade and shipping over time has brought stowaway species into foreign waters. Native marine and estuarine organisms have sometimes suffered when incomers prove to be better hunters. This has happened in the Thames estuary, where the invasive Chinese mitten crab is a fierce predator of local wildlife.

Marine ecosystem overexploitation

Excessive hunting of a particular fish species such as tuna triggers a series of **system impacts**. Sharks can no longer catch enough tuna and their numbers fall too. In contrast, numbers of organisms lower down the food chain might actually increase at first (mackerel numbers could boom with fewer tuna preying on them).

If fishing ceases, the original balance may later be restored. However, overfishing has been known to cause permanent species loss in some cases. North America's cod population has never recovered from **overfishing** in Newfoundland during the 1970s. Some large marine species including the Baiji white dolphin were hunted to extinction during the twentieth century. Humpback whale and great white shark populations are unlikely to ever regain their original sizes.

The implications of fish stock collapse for different societies

There are direct and indirect consequences of overexploitation of marine life, both for our own communities (we may lose access to food sources we value) and other affected communities (who may lose their employment). The worst case scenario is a complete **fish stock collapse**. For example, the collapse of the Newfoundland Grand Banks cod fishery in Canada in 1992 put 40,000 people out of work. This area was one of the world's most productive fishing grounds; however, it was devastated by years of overfishing and incompetent management.

Stakeholders included the fishing industry, government and consumers. Collectively, they managed the ecosystem unsustainably. They increased their catch of fish (a **system output**) far more quickly than the natural replacement rate of young fish being born (a **system input**). A system **threshold** was eventually crossed, which led to the collapse of the entire fish stock. It has never recovered.

In Europe, cod was heavily overfished in the North Sea in the 1980s and 1990s but stringent regulations were imposed on the industry before a point of no return was reached. Now, the cod is recovering slowly. However, the Marine Conservation Society (MCS), which offers consumers advice on seafood consumption, still regards North Sea cod as a species to avoid because it remains at historically low levels. The MCS currently advises consumers also to avoid wild-caught Atlantic salmon and Mediterranean tuna.

The impacts of fish stock collapse can affect many different communities who have become interdependent on account of the fish trade, including indigenous communities (see p. 71). Some people in your own local community may make a living from selling fish (Figure 41), while many others value the continuing availability of fish as part of their diet.

Overfishing is the taking of too many fish or other organisms from the water before they have had time to reproduce and replenish stocks for the next generation.

Knowledge check 17

Why does the growth of emerging economies make it harder to tackle the issue of overfishing?

Figure 41 Neighbourhood fish restaurants may rely on continued supplies of cod and other fish

Sustainable management of marine environments

Widely adopted after the 1992 UN Conference on Environment and Development in Rio, the term **sustainable development** means: 'Meeting the needs of the present without compromising the ability of future generations to meet their own needs'.

Three goals comprise sustainability, or sustainable development (Figure 42).

1 **Economic sustainability:** individuals and communities should have access to a reliable income over time.

2 **Social sustainability:** all individuals should enjoy a reasonable quality of life.

3 **Environmental sustainability:** no lasting damage should be done to the environment; renewable oceanic, terrestrial and atmospheric resources must be managed in ways that guarantee continued use.

For the last of these goals to be met, there must be either a significant reduction in world economic output, or new **technological fixes** that increase resource availability and repair environmental damage. The former may not happen voluntarily; this is because economic growth is the goal of free market economies. The latter may prove expensive to implement on a scale large enough to be effective. Actions intended to support sustainable management of the oceans should, therefore, be assessed with a critical eye.

Exam tip

If an exam question asks you to write about sustainable development, be careful not to write an answer that focuses only on the environment. You need to think about economic and social sustainability too.

Figure 42 A model of sustainable development

Self-study task 10

To what extent do the strategies outlined on these pages operate on a sufficiently large or bold scale to have a significant positive impact?

Many efforts are being made to promote more sustainable management of marine environments. Whether they can collectively safeguard the oceans for future generations and promote long-term global growth and stability remains to be seen. Table 23 provides a round-up of actions and strategies implemented at varying scales, which can be evaluated.

Table 23 Sustainable strategies applied at varying scales by different players

Global actions	■ The **United Nations Food and Agriculture Organization (FAO)** aims to 'ensure long-term conservation and sustainable use of marine living resources in the deep sea and to prevent significant adverse impacts on vulnerable marine ecosystems (VMEs)'. As part of its work, the FAO can designate **marine protected areas (MPAs)** in the high seas. UNCLOS states are expected to follow FAO rules and guidelines. However, many illegal fishing activities still occur because of a lack of any real means of enforcement. ■ World Oceans Day is a 'global day of ocean celebration and collaboration for a better future' held every year on 8 June. This collaboration between the charity The Ocean Project, the United Nations and many other partners raises awareness of ocean issues.
International and national actions (fishing quotas/ limits)	■ The EU's Common Fisheries Policy (CFP) is a set of rules designed to conserve fish stocks in European waters. Catch limits called **total allowable catches (TACs)** are regularly updated for commercial fish stocks using the latest scientific advice on fish stock status. TACs are shared between EU countries in the form of national quotas. Sometimes these quotas are controversial. Many people in the UK fishing industry object to having their catches limited; they hope the UK's 2016 decision to leave the EU may result in reduced 'red tape'. ■ The CFP is controversial because of the way it also forces fishing vessels to discard **by-catch** (dead fish caught unintentionally while fishing for other species).

Table 23 *continued*

Local actions (no-catch and conservation zones)	■ Much of the Firth of Clyde in western Scotland has been overfished. Worse yet, scallop fishing has scoured the sea bed. Fishing vessels collect these shellfish using heavy dredging machinery made up of metal chains and rollers. This equipment destroys maerl, a cousin of coral, which provides an important nursery habitat for cod, plaice and haddock. By the 1990s, fish species had all but vanished in some offshore waters. ■ The community on the island of Arran established an organisation called COAST (Community of Arran Seabed Trust). They successfully lobbied the Scottish government to designate the waters of Lamlash Bay as Scotland's first **no-take zone** (Figure 43). All fishing within the specified area has been banned. But as a result, some fishermen have become unemployed.
Businesses (aquaculture production)	■ Intensively farmed salmon and cod are now raised in caged enclosures along many northern European coastlines. Today, more than 1,200 residents of the Shetland Islands work in the **aquaculture** sector. Shetland Aquaculture is an association made up of nearly 50 producers based in the UK's northern-most islands. It was established in 1984 to provide an alternative to traditional unsustainable fishing methods. Production has expanded from just 50 tonnes of fish to more than 50,000 tonnes. Initially, the focus was salmon, but cod, sea trout, haddock, halibut and mussels are now reared. ■ Globally, aquaculture has grown rapidly since the 1980s (Figure 44). However, fish farming can bring new dangers to marine ecosystems. Outbreaks of parasites and disease are common among caged fish that live in cramped conditions. Hundreds of thousands of salmon escape from North Atlantic farms each year, allowing health risks to spread to wild populations.
Citizens (campaigning and consumption)	■ Individuals can do their bit by making shopping choices that do not support unsustainable fishing. For instance, many shoppers avoid buying tinned tuna that has been caught in nets that may have trapped dolphins too. ■ The chef Hugh Fearnley-Whittingstall leads a campaign called 'Fish Fight', which encourages shoppers to try 'less fashionable' fish such as mackerel instead of popular species such as cod and plaice. The campaign has also drawn public attention to (i) the need to avoid produce sourced from overfished areas and (ii) the 'shameful' waste of discarded fish under the EU Common Fisheries Policy. The Fish Fight campaign can be researched at: www.fishfight.net

Figure 43 Signage at Lamlash Bay informing the public they have entered a no-take zone

Fieldwork

Local people's attitudes and actions towards the consumption of fish could provide a study opportunity. You could carry out a survey in local shops to find out where fish are sourced from.

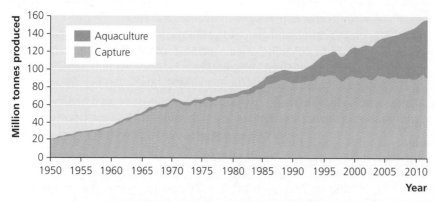

Figure 44 Global fish production, 1950–2012

Summary

- If the oceans are to be used more sustainably as a global commons, then the world's countries and people must agree to conserve and preserve the ocean's ecosystems. Changing attitudes towards whaling over time show it is possible for social norms to change for the better on a global scale.
- Fish stocks must be managed sustainably as a system otherwise there is a risk of total collapse.

However, rising wealth in emerging economies heightens the risk that an overfishing threshold will be crossed.
- A range of strategies and actions have been introduced by different players at varying scales to encourage more sustainable use of oceans; far more needs to be done, however.

■ Managing ocean pollution

Sources, causes and consequences of ocean pollution

The oceans provide us with resources but we also use them as a waste disposal site, both deliberately and accidentally. Oil spillages were examined on p. 59. The following section focuses on (1) plastic materials entering the oceans as a result of **terrestrial runoff** carrying waste from urban streets and landfill sites, and (2) the global pattern of **eutrophic dead-zones** resulting from agricultural pollution.

The plastic pollution problem

Plastic pollution is a problem that has truly 'gone global'. Fragments of plastics washed into the sea by run-off from populated areas have been carried by planetary-scale ocean currents to the remotest corners of the world, including Arctic and Antarctic wilderness areas. The problem has accelerated: more plastic was produced globally in the first decade of the twenty-first century than during the entire twentieth century (the start of which marked the 'birth' of plastic). In 2014, 311 million tonnes of plastics were produced worldwide; this is predicted to rise to over 1,100 million tonnes by 2050. Reasons for the growth in plastic production include:

- the growing use of plastic in everyday life: toothbrushes, credit cards, mobile phones, asthma pumps, Lego bricks, biros, polytunnels and irrigation pipes are all made from plastic

- the cheap commodity boom driven by low wages in developing and emerging economies has fuelled 'throwaway' attitudes on a global scale — if something breaks, it is often cheaper to 'bin' it than to fix it
- the boom in bottled water and other drinks has led to the use of more than 2 million plastic bottles every 5 minutes in the USA. Globally, the figure is far higher. Often, consumption of bottled drinks is driven by 'lifestyle' advertising (given that tap water is perfectly safe to drink in many countries).

Plastic is now believed to constitute 90% of all rubbish floating in the oceans and the UN Environment Programme estimates that every square mile of ocean contains 46,000 pieces of floating plastic. Large areas of Earth's oceans have become particularly polluted with plastic fragments as a result of the operation of **surface gyres**. These are circular currents in the oceans, moving clockwise in the northern hemisphere and anti-clockwise in the southern hemisphere (Figure 45). In the north Pacific Ocean, there is now a floating plastic **garbage patch** which is twice the size of Texas. It is composed of shampoo caps, soap bottles and fragments of plastic bags, as well as much smaller particles called **microbeads** that have been needlessly added by MNCs to toothpaste and shower gels (Greenpeace is campaigning to have microbeads banned globally).

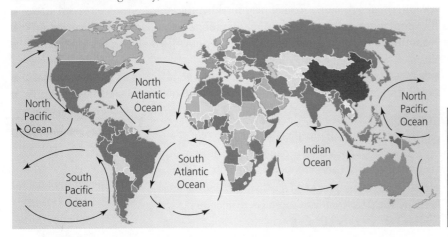
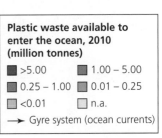

Figure 45 Global pattern of source regions for plastic pollution and gyres that move and trap material

Gyre systems also convey plastic waste to isolated islands and coral atolls far from any pollution source.

- High levels of plastic rubbish have been found on remote Arctic islands more than 1,000 km from the nearest town or village, carried there from polluting countries around the world by ocean currents. Muffin Island is one of the most remote places on the planet, yet plastics from Norway, Spain and the USA litter its beaches. This is a pollution problem that does not respect state boundaries.
- Plastic pollution of the Hawaiian islands, such as Tern Island, has been widely reported by campaigning groups (and provided a stimulus for the recent consumer-led drive to reduce throwaway plastic bag use in the UK).
- Rubber ducks have been washed ashore on once-pristine Alaskan beaches after a floating flotilla of plastic toys was set adrift after a 1992 container ship accident in the Pacific Ocean.

In recent years, scientists have become increasingly concerned with the impacts of plastic pollution on marine species and food webs. Data on sea birds showing the ingestion of plastic waste as being a cause of death first began to appear in the 1950s: 95% of dead fulmars (a common sea bird) washed ashore in Scotland will have some plastic debris in their gut. Worldwide, 260 species of bird and mammal are known to ingest or become entangled in plastic wastes. Discarded red lids from water bottles are a particular problem — in size and colour they mimic the appearance of the krill shrimp eaten by albatross. Autopsies have shown an abundance of red-coloured debris in the gut of dead albatrosses. Large amounts of plastic have also been found in the stomachs of whales. Methods of attempting to deal with the challenge of ocean pollution are discussed on p. 81.

Eutrophication and marine dead zones

Another human impact that brings dire results to estuaries and oceans is **eutrophication**.

- This process occurs when excessive nutrients are added to a body of water. Nitrate fertilisers often get carried by rain runoff from farmland into rivers or over cliff edges into coastal water.
- In contrast to toxic pollutants that kill, the result here is one of over-feeding: the nutrient-enriched waters initially experience a sudden growth in marine life. Tiny organisms flourish, creating an explosion of life called an **algal bloom**.
- For a variety of reasons, the presence of so much algae uses up most of the water's oxygen. Fish and crustacean (crab and prawn) species suffocate in the de-oxygenated water.

Around 20 major **marine dead zones** lie scattered around the world's coastal margins, with Japan and the Gulf of Mexico particularly badly affected. The North Sea is another nitrate hotspot, where lobster populations have been lost because of a lack of oxygen. Some of the worst-affected areas are hub regions for global agribusiness, such as the Gulf of Mexico.

Management of marine waste at varying scales

Positive actions can take place at a range of different geographic scales. These include:

- global conventions (UNCLOS)
- international EU and national rules on waste management
- individual (citizen-led) and non-governmental organisation (NGO) actions

Successful global governance involves a range of complementary actions being taken simultaneously at different geographic scales. By reinforcing one another, these collective actions may, in time, achieve a positive outcome for people and places.

Knowledge check 18

How important is the role of (a) ocean currents, (b) global economic systems and (c) poor governance of waste disposal in the creation of oceanic garbage patches?

Strategies to reduce plastic waste in Earth's oceans

This section uses the management of plastic debris in the oceans as an example of wicked problem — multi-scalar waste management.

The challenge is enormous, as pp. 78–79 explained. Point-source pollution of coastal waters with plastic debris takes place in every country with a coastline: rivers and runoff from urban areas and landfill sites carry vast quantities of waste into the oceans. The volume of plastic waste is growing almost exponentially each decade.

A **wicked problem** is where the issue seems too big in scale for any single action or organisation to make a difference.

Global conventions and their limitations

States bound by UNCLOS rules are not allowed to dump waste deliberately at sea. Instead, terrestrial landfill sites are used for refuse disposal. Many countries also have recycling schemes that intercept plastic waste. So why does so much plastic waste enter the oceans? The problem is land-based discharges of waste that are carried accidently into the marine environment from rivers, estuaries and the coastline. Heavy rainfall flushes plastic litter into sewer systems; overland flows can carry plastic bottle tops and bags from landfill sites into river networks. The vast quantity of ocean plastic is therefore attributable to waste mismanagement, population density on coasts and water cycle movements (also see Figures 49–52 on pp. 89–90).

National and European rules

Some state governments are now using legislation to reduce the use of throwaway plastic that can accidentally pollute the environment.

- Various governments have taken action to ban plastic bags or microbeads. Government restrictions on the use of throwaway plastic bags exist in China and Bangladesh, where the use of thin (<0.025 mm thickness) plastic bags has been prohibited (these small bags also block watercourses and sewers during the monsoon season).
- There has been a 70% reduction in the use of plastic bags since charges were introduced to Wales in 2011 (England followed in 2016).
- The European Commission is currently developing a range of measures that will require all EU Member States to take similar measures to reduce consumption of lightweight plastic bags.
- The USA will ban microbeads from 2017; many global retailers are already removing them voluntarily from their own products.

Awareness-raising and local actions

Global and national actions to tackle ocean waste have been criticised for not doing enough and being hard to enforce. As a result, many citizens and NGOs have joined a growing global campaign to tackle plastic pollution.

- The campaign group Adventure Ecology built a boat called 'Plastiki' made from 12,500 plastic bottles. They sailed it across the Pacific Ocean and through the garbage patch. This caught the eye of the media, raising awareness of the pollution problem. During their voyage in 2010, the expedition crew released videos of the plastic garbage patch onto the internet.
- Many more NGOs — including Greenpeace, The Ocean Cleanup and the Marine Conservation Society — have campaigned on the issue; important foci include the dangers posed by plastic bags, bottle tops and microbeads.

- *Plastic Bag* is a short propaganda film created in 2010 by an international team including American director Ramin Bahrani, Germany's Werner Herzog and members of the Icelandic rock band Sigur Rós.
- Judith and Richard Lang make artworks using plastic recovered from Californian beaches: their work has been displayed widely and raises awareness of the issue.
- There are numerous NGOs dedicated to banning the sale of plastic bottled water in countries where tap water is available.
- The NGO Ocean Cleanup has raised money from its global network of supporters — using an online **crowdfunding** platform — to build a €1.5 million prototype floating barrier made of rubber and polyester which can catch and concentrate debris. Nicknamed 'Boomy McBoomface', it was launched off the coast near The Hague in 2016. The aim is to upscale the model to produce 100-km V-shaped barriers positioned in the Pacific gyre.

Evaluating the success of plastic waste strategies

Many different stakeholders are now working at a range of scales to tackle the problem; working together, they are making progress towards reducing the flow of plastic in some countries. In EU states, government action is being taken at both national and international levels; citizens are educated about the benefits of recycling and local councils provide the necessary recycling facilities.

However, the success of the plastic waste strategies is jeopardised for two important reasons.

1 Plastic use is projected to quadruple globally by 2050, partly on account of rising affluence in emerging economies where waste management strategies are often less robust. Will any actions we take — especially those at just a local scale, such as beach clean-ups — be 'too little, too late'?

2 Most strategies target **new flows** of new plastic waste rather than addressing the **existing stock** of plastic waste that has entered the oceans. Even if the 'Boomy McBoomface' solution works, what will be done with all of the plastic once it is collected? No-one can answer this question.

Perhaps our best hope may lie with the plastic industry itself, which is starting to take action by developing new materials such as biodegradable or even edible plastic. Research has shown that a milk protein called casein could be used to develop an edible, biodegradable packaging film.

Protecting UNECSO marine heritage sites

Local, regional, national and global protection strategies overlap and interact in the management of marine heritage sites designated by UNECSO (United Nations Educational, Scientific and Cultural Organization).

Global governance by UNESCO

Since the 1972 World Heritage Convention, UNESCO has awarded special status and protection to places or regions that have 'outstanding universal value'.

- **Outstanding:** the site should be exceptional. The World Heritage Convention sets out to define the 'geography of the superlative'.

Crowdfunding is raising sums of money for a good cause or innovation by asking a large number of people to donate a small amount each using an online platform.

Knowledge check 19

Why is it important for a range of actions to be carried out at different geographic scales if we are to tackle the plastic waste problem effectively?

Fieldwork

Some local communities have taken action to reduce the use of plastic locally (one such place is Modbury in Devon). An individual investigation could explore local place actions that connect with an important global issue.

- **Universal:** the site should have significance for all people of the world and sites cannot be added to the list because of national importance only.
- **Value:** UNESCO uses a range of criteria to define the 'worth' of a property, such as species richness or uniqueness.

In 2016, 46 marine sites were on the World Heritage List. They cover a vast range of ecosystem types in both tropical and temperate ocean areas. Mangroves, coral reefs and saltmarshes are well represented on a list that also includes:

- the Wadden Sea in the southern North Sea (where more than 10 million birds stop over every year on their way from their breeding areas in Siberia, Canada or Scandinavia)
- Colombia's Malpelo Fauna and Flora Sanctuary (vital for maintaining shark and fish health in the Pacific Ocean)
- the Galápagos Islands Marine Reserve (a global **biodiversity hotspot**)

Case study: Australia's Great Barrier Reef

Australia's spectacular Great Barrier Reef was one of the first marine listings on the World Heritage List. The world's largest coral reef system is home to 30 species of whale and dolphin, 1,625 species of fish, 3,000 species of mollusc and one of the world's most important populations of dugong (sea cow).

Successful protection of the reef has, until now, been achieved through a network of players acting together to reinforce the message that this unique environment requires special protection (Figure 46).

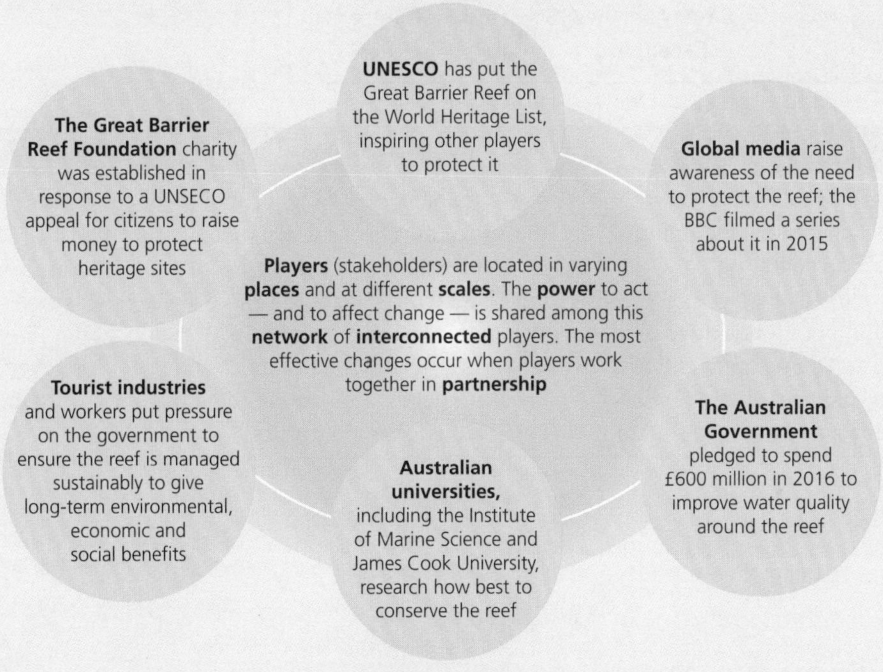

The Great Barrier Reef Foundation charity was established in response to a UNSECO appeal for citizens to raise money to protect heritage sites

UNESCO has put the Great Barrier Reef on the World Heritage List, inspiring other players to protect it

Global media raise awareness of the need to protect the reef; the BBC filmed a series about it in 2015

Players (stakeholders) are located in varying **places** and at different **scales**. The **power** to act — and to affect change — is shared among this **network** of **interconnected** players. The most effective changes occur when players work together in **partnership**

Tourist industries and workers put pressure on the government to ensure the reef is managed sustainably to give long-term environmental, economic and social benefits

Australian universities, including the Institute of Marine Science and James Cook University, research how best to conserve the reef

The Australian Government pledged to spend £600 million in 2016 to improve water quality around the reef

Figure 46 The multi-scalar player network that supports and protects the Great Barrier Reef

Currently, the reef is experiencing the worst crisis in its history. In 2016, sea temperatures in the northern section of the reef rose 2–3°C above the normal peak of about 30°C, because of the strong **El Niño weather system** and a continuing trend towards global warming. A report by James Cook University showed that two-thirds of corals in one part of the reef have died as a result of **coral bleaching** in overly warm water (Figure 47). Many of the players who support the Great Barrier Reef think Australia's government is not doing enough to curb domestic coal production and to tackle climate change. Increasingly, they view climate change campaigning as part of their work in protecting the reef.

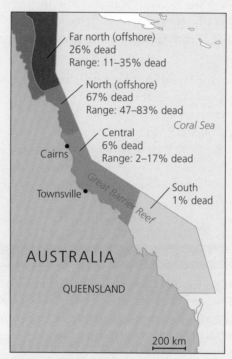

Figure 47 The impact of coral bleaching on the Great Barrier Reef in 2016. The ranges represent upper and lower quartiles (Source: ARC Centre of Excellence for Coral Reef Studies)

El Niño weather system system results from a sustained sea surface temperature anomaly across the central Pacific Ocean that lasts 2–7 years periodically.

Summary

- Ocean pollution is a 'wicked problem'. It has multiple human and physical causes, including rising affluence and ocean currents.
- Strategies have been introduced at varying scales to try to reduce ocean pollution. Global rules are hard to enforce, however, and some local actions achieve little in the 'bigger picture' of things. For real progress to be made, governments must stop mismanaging waste. Recycling needs to become the social norm for all people and societies.
- A range of strategies and actions have been introduced by different players at varying scales to support UNESCO marine sites such as the Great Barrier Reef; climate change requires even greater coordinated action to be taken to protect UNESCO sites in the future.

■ 21st century challenges

Section C of the Global Systems and Global Governance A-level examination addresses 21st Century Challenges. It consists of a single extended response essay, which takes the form of a **synoptic assessment**.

What is a synoptic assessment?

A synoptic assessment tests your ability to draw together elements from across the entire A-level course you are studying. In your A-level course, different topics are studied, usually in a linear order that allows the development of a detailed understanding of each topic in turn. Being synoptic means seeing 'the big picture' and 'thinking like a geographer'. Synopticity therefore involves:

- making links between topics, theories, processes or ideas
- 'joining the dots' to show how different places, societies and case studies may be interdependent or connected with one another
- appreciating the complexity of geographic decision making, especially in relation to 'wicked problems' such as climate change, the world's fossil fuel dependence or new forms of political extremism

Linking themes and concepts

In the examination, you will be expected to spend around 35–40 minutes writing this essay, which is supported by four stimulus illustrations. These will be related to at least two of the core and common themes for WJEC/Eduqas Geography A-level, which are:

- Changing places
- Changing landscapes (coasts/glaciers)
- Tectonic hazards
- Global migration and ocean governance
- Water and carbon cycles

The synoptic assessment is also designed to allow you to make use of appropriate specialised and key geographic concepts. These include the concepts of **causality**, **equilibrium**, **feedback**, **identity**, **inequality**, **interdependence** (see p. 31), **globalisation** (see p. 10), **mitigation and adaptation**, **representation**, **systems and thresholds**, **risk**, **resilience** and **sustainability** (see p. 76). Table 24 explores the use of some of these concepts.

Table 24 Selected specialised concepts used in A-level geography

Key concept: Risk

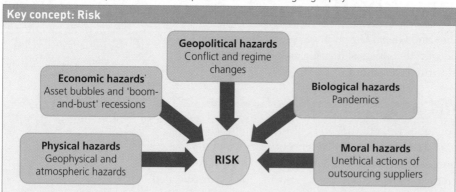

A **risk** is a real or perceived threat against any aspect of social or economic life, or the environment.

- You may be familiar with the concept of risk from studying tectonic hazards. But it can be used in any other geographic context too. For instance, in the context of the study of global systems, physical threats are just one of many risk categories.
- The illustration shows possible risks against a country's people, economy and businesses.

Key concepts: Mitigation and adaptation

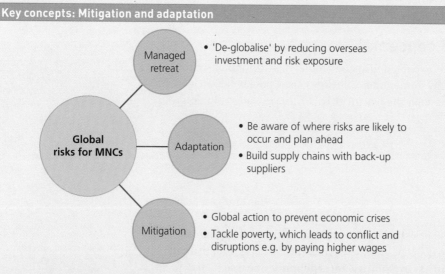

How can we manage risk?

- The illustration shows adaptation and mitigation strategies adopted by large MNCs in relation to operational risks found in their networks.
- **Mitigation** means preventing something from happening, for example reducing greenhouse gas emissions now to stop future global warming, or attempting to end poverty and conflict in those places which generate the largest volumes of migrants.
- **Adaptation** involves dealing with the impacts of something, for example adapting our lifestyles to cope with a warming world, or dealing with the global threat of computer viruses by installing anti-virus software.

→

Table 24 *continued*

Key concept: Resilience

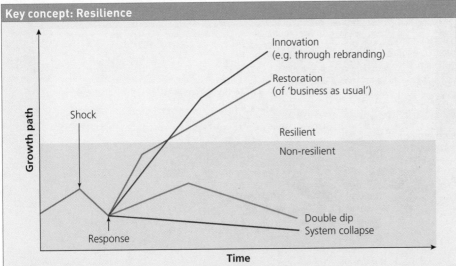

Resilience means having the capacity to leap back or rebound, following a disruption or disaster.

■ The word's roots lie principally in ecology (analysing the self-restorative power of damaged ecosystems).

■ Academics, business leaders and politicians now embrace the word as a catch-all way of characterising the capacity of societies, economies and environments to cope with diverse risks created by development.

■ The illustration shows how the concept of resilience is applied to an economy's response to a financial crisis.

Players, power and perspectives

In addition to making links between topics and using geographic concepts to shape your answer, the synoptic assessment may also allow you to evaluate the importance of different **players** with differing levels of **power**, or to consider an issue from different **perspectives**.

■ **Players:** these are the people, groups, organisations and states that cause geographical issues, suffer the consequences of them, make decisions or attempt to manage them. The term 'player' is used to refer to anyone involved in a geographical issue who can affect — or be affected by — change (see p. 83). Players operate at different interconnected scales, as shown in Figure 48.

■ **Power:** the level of influence of different players over an issue; their ability to determine the outcome of decision making.

■ **Perspectives:** the views of different players, why they are held and how they affect decision making and choice of actions.

Exam tip

It is essential to spend a few minutes planning your synoptic answer. Try to make use of a spider diagram to link your ideas together. Work out the best way to sequence them as paragraphs.

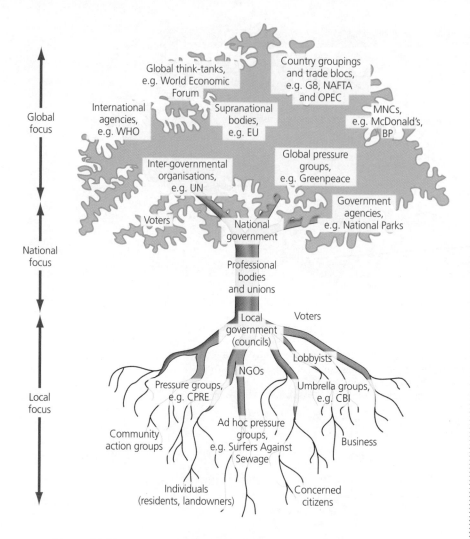

Figure 48 Players involved in geographic issues and decision-making
(Source: Cameron Dunn)

Understanding and applying the assessment objectives

An example of a synoptic essay appears below in self-study task 11. A student answer to this question, with commentary, appears on pp. 101–103. Before reading further, study Table 25. This shows how the synoptic essay is marked: as you can see, there are three separate criteria. It is essential that you understand these, and can produce a structured piece of extended writing that tries to balance the three requirements shown.

Table 25 Assessment objectives and mark allocations for the synoptic assessment

Assessment objective	Marks available WJEC Eduqas	What you need to do
AO3 Use a variety of relevant quantitative and qualitative skills to interpret, analyse and evaluate data and evidence	6 10	Make plenty of use of all four figures and reference them explicitly in your answer ('As Figure 52 shows...'). If possible make connections between the figures to strengthen your argument ('Looked at together, Figure 51 and Figure 52 show...'). A large number of the available marks target your analysis of the figures. A substantial amount of your answer should be devoted to this task.
AO1 Demonstrate knowledge and understanding of places, environments, concepts, processes, interactions and change, at a variety of scales	10 8	You cannot hope to gain full marks by using information from the figures only. You are expected to apply your own knowledge and understanding too. This can be done by explaining processes and introducing concepts not shown in the figures but that are relevant to the task. You can also mention briefly any comparative case studies or supporting evidence that helps you build a case.
AO2 Apply knowledge and understanding in different contexts to interpret, analyse and evaluate geographical information and issues	10 12	The essay is phrased in a way that forces you to evaluate, assess or discuss a range of different issues, themes, views or consequences before arriving at a conclusion. This is the hardest AO to perform for many students. You will need to plan your answer carefully so that you can arrive at a conclusion you feel is justified and defendable.
	Total: 26 Total: 30	

Self-study task 11

Study Figures 49, 50, 51 and 52. Using this information and your own ideas, discuss the view that physical factors are mainly to blame for the accumulation of plastic in Earth's oceans.

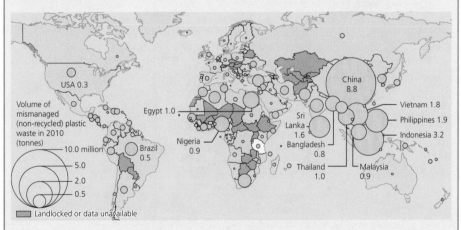

Figure 49 Mismanaged (non-recycled) plastic waste produced by different maritime (coastal) countries, 2010 (Sources: Science, University of Georgia, University of California, Sea Education Association)

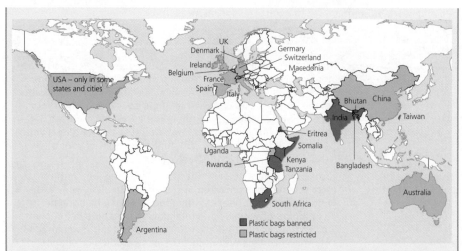

Figure 50 Countries whose government has introduced a plastic bag ban or restrictions, 2015 (Sources: Supreme Creations, National Geographic, BBC, WRAP, ACR Report)

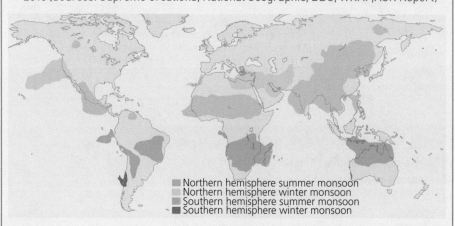

Figure 51 The global pattern of monsoon rainfall capable of generating excess runoff, 2010

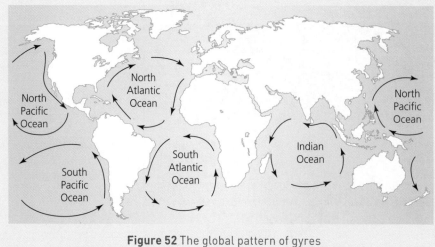

Figure 52 The global pattern of gyres

Questions & Answers

About this section

The questions that follow are typical of the style and structure that you can expect to see in the actual exam papers. Each question is followed by comments, indicated by **ⓔ**, which list the points that a good answer should include and provide a mark scheme. Student responses are then given, with further comments (**ⓔ**) indicating how the answer could be improved and the number of marks that would be awarded. (Note that most of the student answers given here would achieve a high grade overall.) The numbered references included in each student answer indicate the points to which specific comments refer.

Five sample questions are provided. These relate to section B and section C of the WJEC and Eduqas examinations on **Global governance: change and challenges, and 21st century challenges**. Questions 1 and 2 (section B) are accompanied by a figure, which you should use to help you answer the data-stimulus questions that follow. Essay questions 3 and 4 (section B) are worth more marks, and do not make reference to a figure. Finally, question 5 (section C) is the synoptic assessment. This essay is worth between 26 and 30 marks (depending on whether you are a WJEC or Eduqas student) and makes use of four figures.

When the examiners read your work they will use a mark band grid telling them the maximum marks for each assessment objective (AO) (see p. 7). In the example questions that follow, this grid has been placed after each question part.

Question 1

This follows the format used by both WJEC and Eduqas A-level short, structured questions. You have around 11–12 minutes to study Figure 1 and answer both parts of the question (which relates to migration).

(a) Use Figure 1 to analyse the changing importance of global remittances. Include relevant data in your answer.

(5 marks)

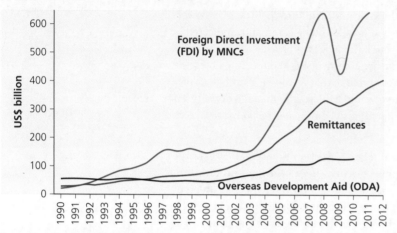

Figure 1 The value of global remittances, foreign direct investment (FDI) by MNCs and overseas development aid (ODA), 1990–2012 (Source: World Development indicators and World Bank estimates)

e Answers might include the following AO3 points:

■ Overall, the value and importance of remittances has risen.
■ Since 2003, FDI has risen steeply, making remittances less important.
■ In 2009, the fall in FDI meant remittances were relatively more important.
■ Remittances fell slightly in 2009 from 320 billion to 300 billion.

Good responses will distinguish between the changing importance of remittances *over time* and also the changing importance of remittances *compared with other financial flows*.

Level mark scheme

Band	Marks	Use Figure 1 to analyse the changing importance of global remittances. Include relevant data in your answer.
3	4–5	Well-developed analysis of the changes shown in remittances over time
		Wide use of evidence to identify changes in the relative importance of remittances
2	2–3	Partial analysis of the changes shown in remittances over time
		Partial or no use of the resource to identify changes in the relative importance of remittances
1	1	Limited statements with no use of evidence
	0	Response not creditworthy or not attempted

Student answer

Remittances are becoming important. They have increased over time from around US$20 billion to over US$400 billion. The rise has been relatively steady but started to increase more rapidly in 2000. There was also a slight fall in 2008 before recovering again in 2010. Therefore remittances have become more important over time.

e **3/5 marks awarded** This is a narrow answer that does no more than describe the remittances trend over time (which is not what the question is asking for). Some data are used, demonstrating the student's quantitative skills satisfactorily. However, there is no reference to the changing importance of remittances in relation to FDI and ODA. Overall, this response is in band 2.

(b) Suggest how the changes in remittances shown in Figure 1 could have affected global economic inequalities. (5 marks)

e Answers might include the following A02 points:

- Remittances are a way of transferring financial resources from core to peripheral regions in the world system.
- They can help reduce inequalities by funding economic and social development. Figure 1 suggests that there is a greater chance of this happening than in the past.
- Some people send money home to their families, which is used to help educate younger siblings.
- Money sent home by migrants is a capital outflow from the host country: this can help to even out inequality globally.

Good answers might apply the concept of interdependency.

Level mark scheme

Band	Marks	Suggest how the remittances shown could have affected global economic inequalities
3	4–5	Well-developed suggestions of how global inequality is affected by remittances
		Wide use of the resource or student's own evidence to support argument
2	2–3	Partial suggestions of how global inequality is affected by remittances
		Partial use of the resource or student's own evidence to support argument
1	1	Limited statements with no use of evidence
	0	Response not creditworthy or not attempted

Student answer

The increase in remittances in Figure 1 would be good news for the places the migrants have come from. Money sent home can be used for a range of purposes, such as education for children or buying a house. Migrants send money home to family members hoping that the money can be used for economic development. Because of the way remittances are rising over time, this means more money is being transferred from rich host countries where migrants live to peripheral countries they have left behind. This means that over time global economic inequalities may be reduced. An example is money sent from the USA to Mexico, which is valued at more than two billion dollars every year.

ⓔ 4/5 marks awarded This is a competent answer that is well focused on the question and addresses explicitly global economic inequalities, using the example of Mexico. Overall, this answer would be in band 3. For full marks, there might be some recognition that the value of remittances is far lower than the annual gross domestic product of the world's richest countries. Also, poor countries have lost their most valued human resources to migration, which may increase inequality (remittances only partially offset this). As a result, remittances may have only a relatively minor impact on global economic inequalities.

Question 2

This follows the format used by both WJEC and Eduqas A-level short, structured questions. You have approximately 12–13 minutes to study Table 1 and answer both parts of the question (which relates to ocean governance).

(a) Use Table 1 to compare piracy trends in different shipping regions. Include relevant data in your answer. (5 marks)

Table 1 Location of piracy incidents involving shipping in southeast Asia, 2010–2014

Location	2010	2011	2012	2013	2014
Indonesia	40	46	81	106	100
Malacca Straits	2	1	2	1	1
Malaysia	18	16	12	9	24
Philippines	5	5	3	3	6
Singapore Straits	3	11	6	9	8
Thailand	2	0	0	0	2
Myanmar (Burma)	0	1	0	0	0
Total	70	80	104	128	141

Source: ICC International Maritime Bureau

ⓔ Answers might include the following A03 points:

- Piracy has risen markedly in Indonesia, from 40 to 100 incidents.
- It has also risen recently from 9 to 23 in Malaysia.
- There has been little change and numbers have stayed low in the Philippines and Singapore.
- In comparison, there is also virtually no activity or change in Malacca, Myanmar and Thailand.

Good answers will make use of comparative language ('whereas', 'compared with').

Level mark scheme

Band	Marks	Use Table 1 to compare piracy trends in different shipping regions. Include relevant data in your answer.
3	4–5	Well-developed comparison of the different regions shown Wide use of the resource as evidence to identify changes
2	2–3	Partial comparison of some of the regions shown Partial use of the resource as evidence to identify changes
1	1	Limited statements with no use of evidence
	0	Response not creditworthy or not attempted

Student answer

Piracy has risen steadily in total since 2010. Overall, there has been an increase from 70 to 141 actual and attempted piracy incidents. The greatest increase has been in Indonesia. The number of incidents has risen from 40 to 100 there. In other areas, the trend is less clear. There are very few incidents in Myanmar and Malacca. In Malaysia, incidents fell between 2010 and 2013 before rising rapidly from 9 to 24 in 2014. Overall, the trends are very different when you compare the different regions.

e **4/5 marks awarded** A sound attempt has been made to compare different regions. The most important trends have been identified and quantified. An explicit comparison is offered in the final sentence. Overall, this deserves a band 3 mark. Note that the first two sentences are not needed because they do not answer the question asked (which is to compare the trends in different regions).

(b) Outline the main security issues which threaten maritime trade, other than piracy. (5 marks)

e Answers might include the following A01 points:

- Security issues other than piracy include oil transit chokepoints.
- The governance of the Suez Canal and the Panama Canal is important.
- Smuggling and people trafficking are criminal activities that could affect legitimate trade.
- Disputes over EEZs and islands can escalate into conflict that might harm shipping.
- Good answers may mention how superpower conflict in the past has disrupted trade (1939–1945).

Level mark scheme

Band	Marks	Outline the main security issues which threaten maritime trade, other than piracy.
3	4–5	Well-developed outlining of a range (at least two) security issues
		Sustained use of supporting evidence to show how trade is affected
2	2–3	Partial outlining of one or two security issues
		Some use of supporting evidence to show how trade is affected
1	1	Limited statements with no use of evidence
	0	Response not creditworthy or not attempted

Student answer

There are many different security issues that threaten maritime trade other than piracy. For example, conflict may occur between two states such as happened between Argentina and the UK over the Falkland Islands in 1982. More recently, China has been increasing its naval power in the South China Sea, which threatens US interests there. The Indonesian government has been blowing up Chinese fishing vessels, claiming that they have entered Indonesia's EEZ. This has created security concerns throughout the South China Sea. Oil chokepoints are another security issue. Because so much oil is moved through the Suez Canal, ships can become delayed. This threatens the energy security of the countries that need the oil.

ⓔ 4/5 marks awarded This response is not always well-focused on how the threats relate to maritime trade. It would have been better if a link had been established between conflict in a region and the risks to commercial shipping that might follow from this. Otherwise, some good points are made, using terminology well. This response might achieve a band 3 mark, however it would not achieve full marks because it is not always focused tightly enough on the question.

Question 3

This follows the format used by WJEC essay questions. You have around 25 minutes to plan and write your essay (which relates to migration). Note that in the WJEC examination, you need to choose between two essay questions (each worth 18 marks). One essay will relate to migration only. The other essay will relate to ocean governance only.

Discuss recent changes in the global pattern of migration. (18 marks)

ⓔ Answers might include the following points:

- Different types of migration, i.e. forced or voluntary.
- Migration from emerging economies to developed countries, and from developing countries to emerging economies, i.e. the economic situation can be complex.
- The recent upswing in numbers of refugees mainly because of the war in Syria. The pattern of this refugee movement consists of large numbers moving to neighbouring countries and smaller numbers heading for the EU.
- Globally, large numbers of people continuing to migrate from rural to urban areas. In some countries, such as China, this is now slowing down.

Good answers may use the specialised geographic concepts of globalisation, interdependency or inequality.

Level mark scheme

Band	Marks	Discuss recent changes in the global pattern of migration.	
		AO1 (10 marks maximum)	AO2 (8 marks maximum)
3	13–18	Accurate and wide-ranging knowledge and understanding of global migration patterns	Well-developed and well-structured discussion of recent changes over time
2	7–12	Some accurate knowledge and understanding of global migration patterns	Partial or unbalanced discussion of recent changes over time
1	1–6	Limited accurate knowledge and understanding of global migration patterns	Limited or no discussion of recent changes over time

Student answer

Migration is a movement of people lasting for at least one year. Migrants move for economic reasons but also when they are forced to flee natural hazards, war, conflict and persecution. People who are forced to move are termed refugees. In recent years, the global pattern of migration has become more complex because of the growth of emerging economies and also refugee flows from the Middle East. a

Turning first to economic migration. In the past, the global economic system was relatively simple. There was a core, which consisted of European and North American countries, and a large global periphery of poorer countries. Many people moved from the periphery to the core. For instance, in the 1950s, post-colonial migrants from ex-British colonial territories moved in large numbers to the UK. b Doctors from India arrived to work in the National Health Service set up after 1945. Other people came from Jamaica to work on the London Underground. A similar pattern can be seen where people from ex-French colonies in north Africa crossed the Mediterranean to work in France.

Since the 1990s, this pattern has become more complex. The rise of the BRIC economies and other emerging countries has meant that people in the world's poorer countries now have a wider choice of relatively affluent states to move to in order to find work. For instance, many people from India have moved to Qatar and to Dubai in order to work in the construction industry there. c The movement can occur the other way around too. Large numbers of Chinese workers have migrated to some of the least developed African countries like Angola and South Sudan where they have found work with Chinese oil and mining companies who have begun to exploit Africa the way Europeans used to do in the 1800s. Global patterns of inequality have become more complicated and so this has changed the pattern of global flows such as migration. d

Not all migration is voluntary. Of the 250 million people not living in the country they were born in, around one-half are refugees e This means they have been forced to flee the country they live in due to war, persecution or a natural disaster. In recent years five million people have fled Syria. Two million are living next door in Turkey. Many more have attempted to gain asylum in Europe. Thousands have died trying to cross the Mediterranean in unsafe boats. They pay large amounts of money to people smugglers who do not have their best interests at heart. The loss of life has been terrible and the newspapers have been full of reports of dead people being washed up on Greek islands and beaches. f

Another refugee movement is people fleeing drought in the Horn of Africa. Many people have fled Somalia for instance and headed south into Kenya. Then they are dependent on shelter provided by the United Nations and the World Health Organization. Refugee camps can provide people with food and water and the shelter they need. However they are not able to offer employment and so the plight of refugees continues. g

> In conclusion, the economic pattern of global migration has become far more complex due to a 'three-speed' world with multiple movements taking place (instead of the old core–periphery system). Refugee movements have always taken place but by their nature the source and destination change as problems are created and solved. Today the out-flow from Syria creates a pattern; in future years there will be different patterns developing around states where new problems have developed. **h**

e **15/18 marks awarded** This a highly competent answer that reaches the middle of band 3. It is well structured and distinguishes between voluntary and forced movement clearly. In places there is excellent supporting evidence, although the focus is not maintained throughout. The conclusion completes the evaluation in a nuanced way, which shows the writer is thinking critically. **a** This introduction is a suitable length: key concepts are established (voluntary and forced migration; the global pattern of migration). **b** There is good use of terminology and evidence here which scores highly for AO1. **c** This shows good contemporary and evidenced understanding of the world. **d** This is an excellent point: the global pattern includes new flows from emerging to least developed countries. **e** A new strand of argument begins here after the paragraph break: the essay is well structured. **f** In this section, the student has lost focus. This reads more like a case study of refugee movements and its consequences than it does a discussion of the global pattern of migration. **g** This also lacks focus on the question. **h** This final point about refugees being a constant element of the global pattern, but the countries where conflict occurs and people flee from changes over time, is excellent.

Question 4

This follows the format used by Eduqas essay questions. You have approximately 20–25 minutes to plan and write your essay (which relates to migration and ocean governance). Note that in the Eduqas examination, you need to choose between two essay questions (each worth 20 marks). Both essays are designed in a way that will draw equally on your knowledge and understanding of migration and ocean governance. You are expected to write about both topics in the single essay you choose.

Discuss the view that barriers to global flows and movements are growing. (Refer to both migration management and ocean governance in your answer.) (20 marks)

e Good answers might include the following points about the governance of migration and oceans.

■ Some countries have strengthened barriers against migration, especially since the terror attacks of 2001, e.g. USA.

■ The UK has voted to leave the EU in order to regain control of migration if possible.

■ However, some parts of the world are moving towards free movement (Africa, South America) and the EU continues with unprecedented free movement.

■ Technology has reduced travel time (time–space compression), which was a barrier to movements.

- Large containers allow commodities to be shipped easily, overcoming the barrier of distance.
- Trade barriers may return and reduce ocean freight movement in the future (e.g. US tariffs).
- Fibre optic cables continue to expand in capacity.
- Piracy and security issues can become barriers to ocean movements.
- Some countries block internet data, e.g. China, North Korea.
- Overall, data flows continue to grow and communication barriers are falling; however, some types of trade and population movements may see increased restrictions in the future.

Level mark scheme

Band	Marks	Discuss the view that barriers to global flows and movements are growing. Refer to both migration management and ocean governance in your answer.	
		AO1 (10 marks maximum)	**AO2 (10 marks maximum)**
3	14–20	Accurate and wide-ranging knowledge and understanding of global flows and movements	Well-developed and balanced discussion of the view that barriers are growing
2	8–13	Some accurate knowledge and understanding of global flows and movements	Partial or unbalanced discussion of the view that barriers are growing
1	1–7	Limited accurate knowledge and understanding of global flows and movements	Limited or no discussion of the view that barriers are growing

Student answer

Globalisation is the lengthening and deepening of connections between people and places on a global scale. There are economic, social, cultural and political dimensions to this process. It is important to recognise this because while barriers may be rising against certain aspects of globalisation in some places, it may not be the case that all aspects of globalisation are being rejected. [a]

Looking first at economic flows, there is plenty of evidence to suggest that barriers to global trade flows have recently increased in some places. This is called protectionism. Back in the 1930s this was a major cause of the Great Depression. Since the Global Financial Crisis of 2008, slower economic growth has led some governments to try and protect their industries in order to save jobs. [b] Many European countries and the USA have been calling for greater import taxes on Chinese steel. When the Port Talbot steel works in Wales were threatened with closure in 2016, 30,000 jobs were put at risk. It is easy to see why governments might be thinking twice about free trade. If imported Chinese steel is taxed more highly then less will be shipped to Europe. [c]

Turning next to migration, the evidence is also unclear as to whether barriers are rising or falling — perspectives differ on this. d Recently the UK voted to leave the European Union on account of many British citizens' desire to reduce migration flows into the country. Many other European countries have seen growing support for anti-immigration movements too, for example France, where Marine Le Pen is leader of the Front National party. In the USA, many citizens are opposed to migration from Mexico and the rest of Latin America. In his presidential campaign, Donald Trump vowed to build a wall along the Mexican border to stop migration. e

However, in other parts of the world there is growing enthusiasm for free movement of people and visa-free travel. The African Union is taking steps to make movement easier for all 54 of its member states. South American countries have also agreed to make temporary residency rights easier to gain. Therefore it is unclear overall whether global migration is on the rise or decline. A record number of 250 million people currently live outside the country they were born in and it seems unlikely that this number will fall any time soon. f

Finally, let us look at global data and information flows carried by undersea cables. These play a role in economic globalisation because services and goods are increasingly traded internationally online. Data flows also contribute to cultural globalisation by sharing music, ideas, languages and other aspects of culture. It is true that internet freedoms are threatened in many places. Iran may soon join China in having its own 'splinternet'. This is an area of the internet that is walled off from the rest of the world. Despite this, total global data flows are at an all-time high. Smart phones and Facebook are barely a decade old. g Their use increases every year as more people are able to afford phones and more people participate in social networks which are global in scale. Capacity also increases as MNCs like Google continue to add new undersea cables to their networks.

In conclusion, the extent to which barriers are rising depends on what global flows and movements you are looking at. f

e **16/20 marks awarded** Overall this a competent answer, which reaches band 3. The application of knowledge and understanding from a range of topic areas here is good. The conclusion is rather short; happily, ongoing evaluation has taken place throughout the essay, which compensates for this. Overall, there is no reason why this answer should not reach the highest mark band. A more substantial conclusion would lift the mark higher.

a This is a clear introduction that deconstructs the key term and establishes a discursive structure. b There is good recall of terminology here. c The ideas are supported by the use of detailed examples. d A new theme is introduced to the discussion here, with opposing perspectives acknowledged from the outset. e This paragraph makes good use of contemporary events and case studies. f Once again, there is ongoing evaluation, grounded in evidence. Strong counter-arguments have been made about migration as part of the evaluation. g A new theme is introduced; again, the material is well-argued, evaluative and well-evidenced. [h] This conclusion is too short. The student also fails to make a judgement as to whether they agree or disagree overall with the statement. One view might be that barriers in global flows and movements are continuing to fall *in general*, thanks to time–space compression, which is a continuing process.

Question 5

This synoptic question follows the format used by both WJEC and Eduqas Section C essay questions. You have approximately 35 minutes to plan and write your essay (which is a synoptic assessment drawing on several ideas from across your geography course). The maximum mark for Eduqas is 30 marks and for WJEC it is 26 marks.

Before attempting this question, look back and study Figures 49, 50, 51 and 52 on pp. 89–90.

Using this information and your own ideas, discuss the view that physical factors are mainly to blame for the accumulation of plastic in Earth's oceans.

(30 marks Eduqas/26 marks WJEC)

ⓔ Good answers might include the following points:

- Runoff carries waste into the oceans; the problem may be worse in monsoon regions with high-intensity water cycle movements (Figure 51).
- Gyre currents trap and carry plastic pollution and lead to its accumulation in particular oceanic regions and coastlines (Figure 52).
- Plastic use is a product of economic development and technology over time. In particular, large amounts of plastic waste now come from emerging economies (Figure 49).
- Some countries have plastic bag bans, including India and China and other countries affected by heavy rainfall and runoff (Figure 50).
- Other uses of plastic could be regulated too and more could be done to recycle. The political failure to do this is the main reason for the problem of pollution.
- A final evaluation might be that the statement is wrong and human factors and not physical factors are primarily to blame: development and wealth create the problem and there is political failure to manage it.

Level mark scheme

Band	Marks WJEC Eduqas	Using this information and your own ideas, discuss the view that physical factors are mainly to blame for the accumulation of plastic in Earth's oceans.
3	20–26 22–30	Mostly accurate knowledge and understanding of a range of factors (AO1)
		Well-developed and structured evaluation of the importance of physical factors (AO2)
		Wide use of the figures as evidence to support arguments (AO3)
2	10–19 12–21	Partial accurate knowledge and understanding of a range of factors (AO1)
		Partial or unbalanced evaluation of the importance of physical factors (AO2)
		Partial use of the figures as evidence to support arguments (AO3)

→

Band	Marks WJEC Eduqas	Using this information and your own ideas, discuss the view that physical factors are mainly to blame for the accumulation of plastic in Earth's oceans.
1	1–9 1–11	Limited accurate knowledge and understanding of a range of factors (AO1)
		Limited or unbalanced evaluation of the importance of physical factors (AO2)
		Limited or no use of evidence from the figures (AO3)
	0	Response not creditworthy or not attempted

Student answer

Are physical factors mainly to blame for the accumulation of plastic in Earth's oceans? This essay will look at both sides of the argument. On the one hand, ocean currents and the movements of the water cycle play an important role in transporting plastic to the oceans and its movement around them inside the large gyre currents. However, there would clearly be no plastic pollution without people. Plastic waste can be controlled, managed and recycled if there is strong government. **a**

Figure 51 shows that large areas of the world are exposed to either the summer or winter monsoon globally. **b** Some of the areas shown are among the most densely populated on Earth. They are also countries where rapid economic growth has resulted in many people becoming part of the global middle class. This consists of people earning around $10.00 a day who can afford consumer items such as bottled water and items which are put in plastic carrier bags when they go shopping. **c**

As Figure 49 shows, China, which is now the world's largest economy, generates nearly 9 million tonnes of mismanaged plastic waste every year. **d** This could mean that the waste is dumped in the streets or in landfill sites. However, when heavy rain occurs this can lead to rapid overland flow. This is because there is insufficient time for water to infiltrate the ground. Instead, it runs over the surface of the land towards the nearest river. This can carry plastic bags and bottle tops into the water system. Eventually, rivers carry this plastic waste to the oceans. As a result there is now a plastic garbage patch in the Pacific Ocean the size of Texas. **e**

This North Pacific garbage patch is growing over time as more and more material is trapped by the gyre current. These currents are shown in Figure 52 and, as you can see, there are five of them. Because of the way they operate, plastic waste can be washed onto the beaches of even unpopulated wilderness areas in Alaska, North Canada and Pacific islands. Dead sea birds have been found in Pacific islands whose stomachs are full of plastic bottle tops. Therefore, clearly, physical factors are playing a major role in leading to the accumulation of plastic in some parts of the oceans where no people are living. **f**

However, just because plastic waste is being produced by affluent societies it does not necessarily follow that the oceans should become polluted. As Figure 49 shows, the level of mismanaged plastic waste in the USA is a fraction of that of China despite the fact that the USA has the world's fourth largest population. g It is also a rich country where many billions of plastic bottles of drink are purchased annually. But as you can see, very little waste is mismanaged and this suggests that if more countries managed their waste effectively and introduced compulsory recycling schemes then there would be far less plastic pollution entering the oceans. h Levels of mismanaged waste are also low in the European Union which has strict rules on waste management and recycling. Far and away the region with the highest levels of mismanaged waste is Asia.

The information in Figure 50 suggests that the countries with the biggest plastic waste problems may now be doing more to try and prevent plastic being washed into rivers and oceans. China and India both have schemes which ban or restrict the use of plastic bags. i Unfortunately, plastic bags are only a small part of the problem, which is probably why these countries still show up as having high levels of mismanaged waste in Figure 49. j But it does offer hope and shows that perhaps the most important factor in the future which will determine whether more plastic accumulates in the oceans will be government attitudes towards waste and the laws that are introduced to prevent it. k

In conclusion, physical factors are not mainly to blame for the accumulation of plastic in Earth's oceans. Rainfall and run-off play an important role in flushing plastic waste into rivers and oceans but really it is a failure of government to manage waste that is the real problem here. It can perhaps be argued that physical factors are largely to blame for the accumulation of plastic in ocean areas that are far away from the places which produce the water. Ultimately, however, there would not be a plastic waste problem were it not for consumers buying and throwing away plastic. Therefore, on balance, I disagree with that statement. l

ⓔ **30/30, 26/26 marks awarded** Overall, this an excellent answer that scores full marks. It balances the assessment objectives of the test well. There are constant references to the four figures, which are used as evidence to back up the arguments made. The answer arrives at a nuanced and persuasive conclusion, which makes a convincing final judgement.

a A succinct and clear start that suggests the essay has been planned carefully first. b AO3 credit for use of the figure. c AO1 credit for application of own knowledge and understanding. d AO3 credit for use of the figure. e AO1 credit for application of own knowledge and understanding. This paragraph mentions infiltration (the water cycle) and the Pacific garbage patch, for instance. f AO2 credit — an evaluation is offered of the relative importance of factors. g AO3 credit for use of the figure. h AO2 credit — an evaluation is offered of the key importance of human (political) factors. i AO3 credit for use of the figure. j AO2 credit — a critical evaluation is offered of what the figure shows. k AO2 credit — an evaluation is offered of the long-term importance of factors. l AO2 credit — a nuanced, evaluative conclusion, which does far more than simply agree or disagree with the statement. There are several reasoned steps of argument here.

Knowledge check answers

1 There is no definitively right or wrong way to write this essay. In order to assess how well you have done overall, reflect on the following three criteria and apply them to your own writing. First, have you provided a clear structure? Have you started a new paragraph when addressing a new aspect of globalisation, such as political globalisation? Second, have you used terminology in your response or have you written in a generalised way? A good answer might refer to ideas and concepts such as diaspora, shrinking world or MNCs, for instance. Third, what quality of evidence have you used? Have you backed up your ideas with precise and detailed information? For example, you might have identified by name some countries that the fruit and vegetables you eat are sourced from.

2 See the table below.

Flow	Links with cultural change, development and democracy
Food	Trade in foods can help countries develop: many African states still depend on agricultural exports for much of their GDP. Trade in food is also linked with cultural change: many foods are now imported routinely into the UK that were not in the past, and British diets have changed markedly over time as a result.
Money	Flows of investment are important for economic development — China provides a striking example of a country where rapid development has taken place on account of productive investment by foreign companies such as Apple. Some people view wealth creation as part of a broader development process that may lead to greater democracy.
Migrants	Migrants send home remittances that can help the development of the source nation. For example, US$10 billion returns to Mexico annually. Migrants also bring cultural changes with them — Asian and Caribbean migration to the UK has brought many cultural changes to the UK, including music and food.
Ideas	Migrants may bring ideas with them or return home eventually with new ideas. Over millennia, migration has led to the spread and mixing of different ideas, languages and religions.

3 The shrinking world process can be attributed to transport technology and also the internet and global media. Themes you could explore include the way faster transport such as aeroplanes has made it quicker and easier to migrate compared with steam ship travel 100 years ago. Another theme is the way social networks and apps allow potential migrants to discover information about faraway places. This may lead to more people wanting to make the journey. In order to make an assessment about which technology has had the greatest impact, think about different timescales and spatial scales. Migration has been happening for thousands of years whereas the internet has only existed for decades. It might therefore be sensible to conclude that steam ships in the nineteenth and twentieth centuries made the greatest impact when one considers how many people travelled originally from Europe to the USA or Canada. Migration from the Caribbean to the UK in the 1950s was also reliant on shipping.

4 Your answer will depend on the location of your school. Students in schools in some London Boroughs may find that the majority of their friends belong to an Asian, European or Caribbean diaspora community. Many famous people belong to a diaspora community. Good places to look for evidence of this include: premier football teams; the cast of television soap *Eastenders*; and well-known musical acts featuring in the MOBO (British Black music) awards.

5 See the table below for suggestions. Perspectives may differ on what is the most important criterion for superpower status. One view might be that economic power matters most, as without it there may be little funding available for military spending or to support media capable of projecting a country's soft power.

Nigeria	Major regional power because of status of Lagos as economic hub; some global reach through 'Nollywood' film industry; a major global oil player
South Africa	Major regional power because of strength of economy; some global reach through its sports and culture
India	Emerging global power because of population size (1.3 billion) and strength of economy, including MNCs (Tata); strong global cultural reach (Bollywood, Indian music and food)
Russia	Military global superpower and strong regional power (mainly exerted over eastern Europe through control of gas supplies); strong cultural influences (e.g. Bolshoi)
Brazil	Top ten world economy and major regional power; global cultural reach through culture, music and sports

→

Japan	Economic global superpower and strong regional power with a significant navy; global cultural and technological influence, e.g. Pokémon and Sony
Saudi Arabia	Major economic influence because of dominant role within OPEC oil cartel

6 One argument might be that superpower countries rely on innovation and technology to keep moving their economies forward. Therefore it is important to attract highly skilled technological experts such as leading physicists or computer programmers. The USA benefited greatly from the arrival of many scientists who fled Europe in the 1930s and 1940s, for example Albert Einstein. Or we can consider how essential migrant labour was for Victorian Britain: large numbers of Irish workers helped build the London Underground for instance. More recently, many new Middle Eastern powers such as Qatar and UAE have been dependent on Indian migrants working in construction industries. However, Japan may be something of an anomaly. It has the world's fourth largest economy yet remains relatively closed to permanent migration.

7 This presents you with an opportunity to think and write critically about the key geography concept of interdependency. In both of the examples shown, the host countries have a far higher GDP per capita than the source countries. The host countries are benefiting from the labour of millions of people, many of whom worked long hours for low pay. The absence of large numbers of young and motivated workers from India and the Philippines may mean that both countries have lost valuable human resources. The extent to which this is compensated for by remittance flows is debatable. However, many economists believe that remittance flows do not always compensate for the loss of large numbers of young workers who have been trained and educated at the source country's expense.

8 Your answer depends upon where you live, obviously. Several cities and regions where large numbers of people are non-UK born show positive attitudes towards immigration. Perhaps this is because large numbers of respondents to the questionnaire were in fact migrants! It is also due to the fact that UK-born people in these communities have attended multicultural and multi-faith schools and, based on their own experiences, genuinely believe that immigration has enriched cultural life. Other things to consider are the views of people who live close to areas where there has been high immigration but who have not experienced it personally. Be careful when answering this question: the focus is on attitudes to cultural and not economic life. Therefore people's responses have not been affected by issues surrounding job availability.

9 The examples used in the text go a long way towards showing the interlinked factors affecting migration.

It is noteworthy that many analysts believe that climate change may have exacerbated drought and population movements in states and regions where there is now conflict, including Syria and Sudan. Land grabs can result in pastoral farmers being forced to move towards areas of marginal land affected by drought. Migrant groups may also find they are competing with settled farmers for the same land resources. Therefore, the different causal factors frequently become interlinked in some of the world's poorest countries and regions.

10 The examples used in the text show that island ownership is an important geopolitical strategy for countries wishing to increase the area of water they can claim territorial rights over. Examples include the UK's claim over the Falkland Islands, and China's island claims in the South China Sea. A small island can generate a large circular EEZ with a diameter of 400 km. If these waters are known to contain significant biotic resources, and the seabed is a source of important abiotic resources, then it is easy to see why ownership of island territories becomes important.

11 A rough estimate using the figure is that three-quarters of intra-regional trade flows link together world regions that are not connected by land. The only significant flows taking place over land implied by the diagram are between Asia and western Europe, and between eastern Europe and western Europe. Some of the flows shown may not take place as the crow flies (which is how the diagram portrays the flows). Therefore we should treat any estimate made from this diagram with a degree of caution. The largest trans-ocean flows — if this is what they are — appear to take place between Asia and North America, and western Europe and North America. This reflects the fact that the world's largest consumer markets are found in developed countries in North America and western Europe, and also in Japan. Increasingly, China is an important market as well as a production site.

12 Figure 29 shows a huge expansion in container movements between 2000 and 2014. The number has risen from 200 to around 800 container movements. This represents a quadrupling of trade flows potentially (although we cannot know that all containers are filled fully). There was a brief downturn in movements between 2008 and 2009. This would be because of the global financial crisis that saw the UK and other countries briefly entering recession, leading to reduced demand for goods among the population at large. Since 2010, recovery has taken place although, as you can see, growth is slower than it was in the run up to the crisis.

13 Different categories of risk include geopolitical threats such as conflict and terrorism, or troop movements. Large companies might want to avoid areas where their operations or products might be attacked, appropriated, or where their staff might be kidnapped and held hostage. Another risk is the moral hazard associated with exploitation of workers and poor health and safety standards in the workplace. Some companies stopped using Bangladesh as an offshoring and outsourcing site following the deaths of workers in a well-publicised factory collapse. Apple now looks closely at its supply chain for evidence of worker exploitation that could give the company a bad reputation. Finally, natural hazard risks remain a major consideration. The Japanese tsunami of 2011 is a good example of how many companies realised their vulnerability to supply chain interruption.

14 The study of plate tectonics introduces students to the idea that there are two types of crust: continental crust and ocean crust. The continental shelf is geologically part of the continental crust and may extend a long distance beyond the country's coastal margin. Because the continental shelf is geologically part of the land mass that makes up a country, there is a strong case to be made that it is in fact recognised legally as being part of that state. This right is recognised under the United Nations Convention on the Law of the Sea.

15 Geopolitical tensions and conflicts arise first because of the way coastlines of different countries often run close to each other. A good example of this is the South China Sea, where parts of Southern China lie close to Vietnam's coastline. As a result, the areas of water extending 200 nautical miles from each country overlap. This requires arbitration in order to decide where the boundary line is drawn. China's claim over the Spratly Islands has resulted in contested claims over ocean that the Philippines claims are its own territorial waters. Geography has led to the *potential* for tension over territorial claims; but population growth and rising affluence have *heightened* tension further. The resource needs of states such as China and the Philippines have increased greatly in recent decades. This has raised the stakes on ocean resource ownership further.

16 The exceptionalism of Japan, Norway and the Faroe Islands reveals a weakness in the argument that global governance can lead to the successful adoption of universal norms. There are nearly 200 states, each of which has its own unique culture or cultures (some states are composed of multiple countries, e.g. the UK). Universal laws and agreements can be hard to reach because of the diversity of different national viewpoints and perspectives. For the countries mentioned, whale hunting remains an important part of their culture even though the majority of countries now oppose it. There are many other examples of global agreements failing to achieve a consensus. Not all

countries signed the Paris Agreement on climate change. Difficulties have arisen when the United Nations has attempted to impose sanctions or approve military action against countries that have broken international laws. This is because other countries often support the lawbreaking state for other political reasons.

17 In recent years, more than 1 billion people living in emerging economies have escaped poverty. In China, India, Indonesia and the Philippines, among others, a further 1 billion people are predicted to achieve middle-class status by 2025. Rising affluence on this scale inevitably puts enormous pressure on the world's resources, including marine ecosystems. While subsistence farmers may have a largely vegetarian diet, the adoption of a meat- and dairy-rich diet generally goes hand in hand with a society's climb up the income ladder. Therefore it becomes harder year on year to manage fish stocks sustainably and to prevent overfishing.

18 All three factors play an important role in the creation of garbage patches. The concentration of waste in large areas of ocean is due to gyre currents: they have a circular motion that concentrates material in particular regions of ocean. However, the basic fact that there is a large amount of plastic waste flowing into Earth's oceans is attributable to human factors. Most important is the global economic system, which has resulted in the mass use of plastics in a growing number of middle- and high-income nations. Economies of scale and efficient production systems have made plastic a cheap, throwaway substance. The challenge for governments is great: the logistics of collecting, containing and recycling plastic are immense. Although governments everywhere could do more to manage plastic waste, the accelerating rate at which it is used makes this a challenge that is hard to tackle.

19 The challenge is so great that it can only be tackled by a range of different stakeholders/players acting together. There are also two different problems to tackle: first, we need to reduce the flow of new plastic into the oceans and, second, we need to take measures to remove and dispose of the huge stores of plastic that have already entered the oceans and which will not disappear anytime soon because of the non-biodegradable nature of plastic. Therefore some actions need to focus on local refuse collection and measures that encourage recycling or reduce the use of throwaway plastic. These may be most effective if reinforced at a national level by government or intergovernmental laws. At the same time, other actions are needed to clean up the mess that is already there.

Self-study task answers

1 South Asia, 1.4 million; East Asia, 0.8 million; Southeast Asia, 0.6 million

2 The smallest points have a value of approximately 5% or less. The highest values are found in some London Boroughs with a maximum of 50%. Therefore the range is 45%. This is calculated by subtracting the minimum value from the maximum value in the data set.

3 A good example is Democratic Republic of the Congo. This large country in Central Africa is shown to contain literally hundreds of ethnic borders. Moreover, some homelands have been split in two so that half the people live in DRC and the rest reside in a neighbouring state, for example Uganda. Many challenges arise. The most obvious is the political challenge that develops when a divided group of people wish to be reunited within a single modern state with their own sovereignty; another challenge is the sheer ethnic diversity that results for some states in terms of the number of different languages spoken and religions. This creates a challenge for state identity-building and community cohesion.

4 Figure 19 shows that the highest numbers of refugees are found in the states immediately adjacent to Syria. These are our Jordan, Lebanon and Turkey. The highest number is found in Turkey — 2,503,550. Further away, the numbers of refugees are lower, for instance in Libya, Spain and Switzerland. An anomaly appears to be Sweden. Although it is further away from Syria than almost any other state, Sweden has received 102,870 asylum applications.

5 There is great variability. Germany has received more than twice as many applications as any other European state. The USA has also received a high number — 121,160. Although Germany and the USA have had a high number of applicants they have only accepted 42% and 30%, respectively. The highest percentage accepted is in Sweden, where 77% of 75,090 people have had their claim accepted. There is also great variation in the source regions for the asylum applications made in each country. European countries have received large numbers of applicants from Syria and Serbia, whereas the USA receives applications from Mexico and China. Within Europe the pattern varies. Half of Hungary's applicants come from Serbia whereas 40% of Sweden's applications come from Syria. Finally, the treatment of asylum applicants varies from country to country. Britain has the longest wait — 12 months, whereas Sweden will permit people to work immediately. Germany offers the most generous benefits, followed by Sweden. Hungary offers the smallest state benefit of just €86.00. These variations reflect the geography of the countries:

European states are relatively easy to reach from Syria and Serbia whereas the USA is easy to reach from Mexico. Some countries receive a high number of applications because asylum seekers believe they will be welcome there. The variations in treatment of the asylum seekers may reflect the affluence of the country, the cost of living and the attitudes of the government and the voters who have elected it.

6 Between 2009 and 2010 the number of attacks increased, peaking at 688. Since then, it has decreased, although the trend is uneven. Numbers rose slightly in 2014 before falling again in 2015 to 306. During this time, the pattern has varied. During the first three years, East Africa was the main location of piracy. For example, in 2011 nearly 400 out of 679 attacks were near East Africa. Since then, Southeast Asia has become the main piracy hotspot.

7 The five largest flows are as follows:

Asia to North America	US$600 billion
Asia to western Europe	US$250 billion
Western Europe to North America	US$250 billion
Western Europe to Asia	US$100 billion
Gulf states to Asia	US$75 billion

8 The answer is 40,000 gigabits per second. This is a complex resource and needs to be studied carefully. Do not rush data analysis tasks. Easy mistakes to make include wrongly including the data for flows within Europe itself, or writing '40' instead of '40,000'. The high value is because this is a wealthy region and incomes are high enough for people to afford smartphones, tablets and laptops. Therefore there is a great deal of online activity such as social networking and online shopping. There is also a great deal of activity in service and quaternary industries, including research by universities that are networked with universities in other world regions.

9 On the one hand, this model suggests that over time environmental impacts of human activity may reduce after having initially increased. Many countries have improved demonstrably their environmental record in recent years and there is a strong record of global governance under the auspices of the United Nations aimed at helping the planet. Therefore it could be that during the twenty-first century we will see a reduction in the harm done to the oceans and its ecosystems. During the twentieth century, action was taken effectively to reduce whaling and this offers some hope for the future. On the other hand, population and affluence continue to rise, which is increasing humanity's ecological footprint beyond a threshold where irreversible damage may occur. It may be far harder to remove plastic from the oceans and protect a wide range of disappearing marine species than it was to protect a single type of animal, e.g. whales.

10 There are many approaches to answering this question. For an effective response, check that you have done the following:
- Refer to all of the strategies outlined on the relevant pages.
- Address the difference in meaning between large scale and bold scale. For instance, the action in Lamlash Bay was at a very small scale. How feasible is it to introduce a no-take zone on a far larger scale and enforce it effectively?

- Provide a balanced answer that praises the scale of some strategies and activities, such as aquaculture growth, in addition to identifying less significant strategies.

11 See pp. 102–103 for a student answer to this question, with accompanying comments.

Note: page numbers in **bold** indicate defined terms.

Index

Index